LIFE SCIENCES MISCELLANEOUS PUBLICATIONS
ROYAL ONTARIO MUSEUM

D. W. NAGORSEN
R. L. PETERSON

Mammal Collectors' Manual

A Guide for Collecting, Documenting, and Preparing Mammal Specimens for Scientific Research

Publication date: 6 June 1980
ISBN 0-88854-255-0
ISSN 0082-5093

ROYAL ONTARIO MUSEUM PUBLICATIONS IN LIFE SCIENCES

The Royal Ontario Museum publishes three series in the Life Sciences:

LIFE SCIENCES CONTRIBUTIONS, a numbered series of original scientific publications including monographic works.

LIFE SCIENCES OCCASIONAL PAPERS, a numbered series of original scientific publications, primarily short and usually of taxonomic significance.

LIFE SCIENCES MISCELLANEOUS PUBLICATIONS, an unnumbered series of publications of varied subject matter and format.

All manuscripts considered for publication are subject to the scrutiny and editorial policies of the Life Sciences Editorial Board, and to review by persons outside the Museum staff who are authorities in the particular field involved.

D. W. NAGORSEN is a Curatorial Assistant in the Department of Mammalogy, Royal Ontario Museum.
R. L. PETERSON is Curator-in-charge in the Department of Mammalogy, Royal Ontario Museum, and Professor in the Department of Zoology, University of Toronto.

Cover: A hoary bat *(Lasiurus cinereus)* and a deer mouse *(Peromyscus maniculatus)*

Canadian Cataloguing in Publication Data

Nagorsen, David W.
Mammal collectors' manual

(Life sciences miscellaneous publications ISSN 0082-5093)

Bibliography: p.

ISBN 0-88854-255-0 pa.

1. Mammals — Collection and preservation — Handbooks, manuals, etc. I. Peterson, Randolph L., 1920– II. Royal Ontario Museum. III. Title. IV. Series: Life sciences miscellaneous publication.

QL708.4.N34 579'.02'02 C80-094365-1
QL1.T6539

100 Queen's Park, Toronto, Canada M5S 2C6
PRINTED AND BOUND IN CANADA AT THE ALGER PRESS

Contents

Mammal Collectors' Manual

1. Introduction

In 1965, one of the authors (R.L.P.) prepared a pamphlet *Collecting Bat Specimens for Scientific Purposes,* in which procedures were outlined for collecting, measuring, preparing, and shipping specimens. The usefulness of that leaflet prompted us to expand the concept into a more detailed book that includes most mammals. Our intention here is not to encourage indiscriminate collecting. Rather the purpose of the manual is to provide a guide that will serve as a set of standards for anyone collecting mammals for scientific specimens. By following the guidelines in this manual, specimens and associated data should have maximum value for research. In addition to museum biologists and mammalogy students, the guide should be useful to ecologists, biologists involved in environmental impact studies, parasitologists, cytogeneticists, and other laboratory scientists who may find it necessary to prepare voucher specimens for their research. Naturalists and wildlife biologists wanting to prepare museum specimens from rare or unusual mammals obtained from hunters, trappers, or road kills should also find the manual helpful.

Several general guides already exist for the mammal collector. These include Hall (1962), Anderson (1965), Setzer (1968), Giles (1971), and DeBlase and Martin (1974). However, several of these publications are now out of date and do not describe the newer techniques available. Also, the kinds of field data that should be recorded may be given only cursory coverage. A number of specialized papers (see bibliography) dealing with collecting techniques for specific mammals or field data (reproductive data, locality descriptions) exist in various scientific journals. However, the average collector may be unaware of these publications. Therefore, we have attempted to produce an up-to-date guide on the methods for collecting, preparing, and documenting mammal specimens by combining methods derived from our own field experience and from the published sources listed in the bibliography.

The recording of field data is strongly emphasized in the manual. In 1972 the Department of Mammalogy at the ROM initiated a computerized cataloguing system for storing and retrieving specimen data. With this system, it is now possible to process a large quantity of data for each specimen. In addition to standard measurements, information on weight, sex, age, date, reproductive data, habitat description, and a precise locality description, including latitude and longitude, are stored on magnetic tape. Other institutions with major mammal collections have also begun to use similar computer systems and it is possible that in the future most museum catalogue records may be stored in a central data bank. To utilize fully the potential of these cataloguing systems, collectors should provide the maximum amount of data for specimens.

With an increasing concern for the conservation of mammalian species and the additional restrictions being placed on collectors, it is most important that a reasonable and responsible collecting policy be followed. Collecting ethics and collecting laws are discussed in section 2. The recent proliferation of import/export regulations for scientific specimens is another area of concern for collectors. Canadian and US import/export regulations are discussed in some detail in section 7. Collectors are urged to read these two sections carefully before collecting specimens.

2. *Collecting Policy*

2.1 Collecting Laws

In recent years there has been a great increase in the number of collecting laws and endangered species acts that directly affect the scientific collector. These laws may be complex and ambiguous and obtaining the necessary permits for collecting in an area may involve considerable bureaucracy. Nevertheless, the collector has an obligation to learn and comply with these laws and regulations. It is essential that permits be obtained prior to any field collecting. Mammals that are protected under endangered species legislation should not be disturbed or collected except under special permit.

Canada

In Canada, a Scientific Collector's Licence issued by the various provincial and territorial governments is needed. Fur-bearers and game species are usually regulated under provincial Game Acts and Regulations and special permission may be necessary for collecting these species. Additional permits may be needed to work in a provincial park. In Ontario, for example, collectors who plan to work in provincial parks must have their research proposal approved by the District Manager and the Director of the Parks Branch, Ministry of Natural Resources. Permission from the federal government is required for collecting in national parks. Some provinces and territories also require permits for salvaging dead mammals or parts thereof (carcasses, bones, shed antlers). For more information, consult the appropriate provincial or territorial governments.

Although Canada has no federal endangered species act, Canada has signed the Convention on International Trade in Endangered Species (see section 7.2). Ontario and New Brunswick have passed provincial endangered species legislation.

United States

The US laws are complex and involve state and federal authorities. Generally, wildlife is regulated by the state governments and the collector should contact the appropriate state governmental agency for permits. To collect scientific specimens in a national refuge or in a national park, a permit issued by the Refuge Manager or the

Superintendent of the national park is necessary. Permits to collect in national parks are issued only to persons officially representing reputable scientific institutions and annual reports are required. Although state and federal requirements for scientific collecting must be met, special permits are not needed for collecting in national forest systems. Collectors, however, are requested to contact the local Forest Service District Ranger before initiating any fieldwork.

Marine mammals are covered under the Marine Mammal Protection Act. You may take and process marine mammals and parts thereof (bones, teeth, ivory) only under permit. Walruses (*Odobenus rosmarus*), sea otters (*Enhydra lutris*), polar bears (*Ursus maritimus*), and manatees (*Trichechus* sp.) are under the jurisdiction of the Director of the US Fish and Wildlife Service. Cetaceans and all pinnipeds (except walruses) are under the authority of the Director of the National Marine Fisheries Service.

Collectors working in the US should be aware that endangered and threatened species are protected at both the federal and state levels. The federal Endangered Species Act that took effect in 1977 prohibits the taking and capture of all listed mammals as well as the import and export of these species. Species protected under the Act are listed as either ''endangered'' or ''threatened''. The various prohibitions of the Act apply to live or dead mammals and their parts or products. If you plan to salvage or utilize dead mammals listed in the Act, you must have a permit issued by the Director of the US Fish and Wildlife Service, or if endangered marine mammals, from the Director of National Marine Fisheries Service. For further information, contact the Federal Wildlife Permit Office, US Fish and Wildlife Service, Washington, DC 20240. Some states have also passed endangered species legislation and mammals protected under state law may or may not be the same as those listed under the federal Endangered Species Act. Information on these state laws can be obtained from the appropriate state governmental authorities such as conservation or wildlife departments (see McGaugh and Genoways 1976).

OTHER COUNTRIES

Many other countries have also passed endangered species legislation and regulations for the collecting of scientific specimens. Collectors planning to work abroad should contact the governmental agencies in these countries well in advance of any field trip. It is essential that one understand the regulations in these countries and obtain the necessary permits before any field collecting. Moreover, export permits may be required in some countries for transporting specimens out of these countries.

The collector should also be aware of the Convention on International Trade in Endangered Species. Even if species listed in the Convention are not protected in the country where the collector is working, it may be impossible to import specimens of species listed in the Convention into North America without permits (see section 7).

2.2 Firearms

In recent years there has been a growing anti-hunting sentiment and collectors therefore are urged to use discretion when collecting with guns. Handguns are strictly controlled in Canada and permits issued by police departments are required. In some provinces municipal or provincial police departments may issue permits; in other provinces, the Yukon, and the Northwest Territories, permits are issued by the Royal Canadian Mounted Police (RCMP). In the US, firearms are generally regulated by the various state governments and collectors should consult their state or local police departments for information. When planning to use firearms in foreign countries, investigate thoroughly the various laws in these countries pertaining to firearms and ammunition. These laws include customs regulations covering the importation of guns and ammunition.

2.3 Collecting Ethics

The possession of a valid collecting permit does not give the collector the right to use irresponsible collecting methods. Specimens should be collected in the most humane method possible and any damage or destruction to the local biota or collecting sites must be prevented. Indiscriminate collecting of excessive numbers is discouraged, particularly in areas where large numbers may concentrate. When possible, take mammals alive and once the required sample has been collected, release the remaining mammals unharmed. Obviously with techniques such as snap trapping this procedure is impossible to follow. Specimens acquired for systematic collections should be carefully prepared and thoroughly documented using the standards outlined in this manual.

Systematic research on a species requires a statistically adequate number of specimens from various localities. Because some species are sexually dimorphic (e.g., one sex may be consistently larger than the other), a representative series should contain samples of both sexes. For a given locality, 10 to 15 adults of each sex are usually an adequate number for a species. A few young animals in each sample may be useful for studying growth and age variation.

To study geographic variation in a species, representative samples from various localities throughout the geographic range of the species are required. Distance between collecting sites is a function of habitat diversity in a given area and in regions with homogeneous biomes and habitats (e.g., the boreal forest in northern Canada or the tropical rain forest in the Amazon basin) localities may be 15 to 160 km (10–100 miles) apart. But in regions that support a diversity of biomes and habitats (e.g., mountainous regions or river systems), collecting sites may need to be close together, 8 km (5 miles) or less.

3. Methods for Collecting Mammals

3.1 Bats

A. Equipment

MIST NETS

Mist nets or bird-banders' nets (Bleitz Wildlife Foundation, 5334 Hollywood Boulevard, Hollywood, California, USA) are effective for capturing bats alive. The nets consist of a fine nylon mesh (50 or 70 denier) thread with mesh consisting of 36 mm (1.5 inch) squares fitted on a string frame that divides the net into panels. Loops of cord at the end of each panel hold the net on supports such as long slender poles of bamboo cane (Fig. 1).

In tropical regions a machete is useful to cut and trim suitable poles and to clear vegetation. To set the net, secure one pole in the ground. Usually the pole can be driven into the ground by repeated jabbing and twisting, however, you may have to provide additional support by piling rocks against the base of the pole or by attaching guys of cord to nearby trees or stakes. Place the loops of one end of the net in proper sequence over the pole and unfold the net, keeping it taut and off the ground. When the net is completely unfolded, erect the second pole and slip the loops over it. Once on the poles, the net is opened by spacing the panel strings at intervals along the pole. It is important that the net be taut enough to prevent sagging but loose enough to provide adequate pockets to stop bats from bouncing off the net. Remove any leaves, sticks, or insects that may become entangled in the net. Although the poles and net should be set and adjusted before dark, the net should not be opened on the poles until the collector intends to use it. This will prevent the capture of birds that are frequently active several hours before sunset. It is important to check nets frequently and regularly and to remove netted bats, as large bats can cause extensive damage by chewing themselves free. To catch species that are adept at freeing themselves without entanglement, it is necessary to actually stand by the net and remove bats as they hit the net. This is particularly true for some of the small, African free-tailed bats (Molossidae). Constant attention is also necessary in agricultural areas where domestic animals may wander into untended mist nets.

Equipment required to attend nets includes: a headlight with spare batteries and bulbs, a flashlight, collecting bags, gloves, nylon fishing line to repair broken shelf

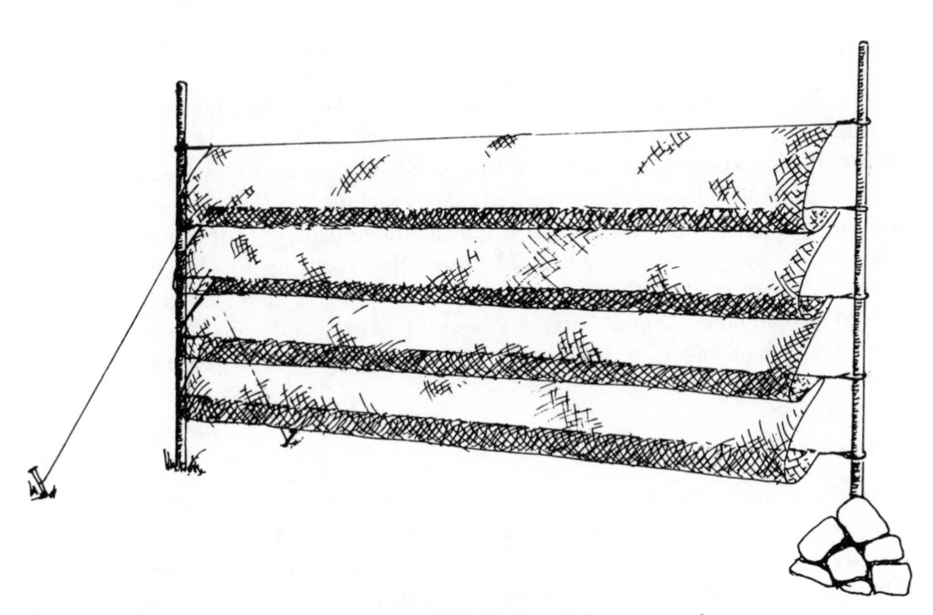

Fig. 1 A four-panel mist net (12.5 m × 2 m; 42 ft × 7 ft) set on poles.

strings, cord or rope for guy lines, and a knife or scissors for cutting badly entangled bats from the net.

To remove a bat from the net, determine first the side from which it entered. With the bat held firmly in the gloved hand, use the ungloved hand to remove the bat from the open side of the pocket, beginning with the head. The best method is to remove netting from the bat's mouth first and then work back, freeing the wings and feet. Once removed from the net, bats can be kept alive in cloth collecting bags about 35 cm × 20 cm (14 inches × 18 inches). Use several bags and keep the larger species separate from the smaller ones. If bats are not prepared immediately as specimens, they can be kept alive overnight in these bags. To increase the chance of survival, keep the bags in a cool, well-ventilated place.

Nets should be taken down before dawn to avoid catching birds and diurnal insects. First, remove all leaves or sticks, then slide the string loops together on the poles to close the net. In areas where theft is not a problem, nets may be closed and left on the poles during the day. When the net is to be removed from the poles, tie a piece of string to the top loop of each end and pass this string through all the loops to keep them in proper sequence. Then remove the net from one pole and walk towards the other pole collecting and folding the net in your hands. The net can then be doubled or redoubled into a compact bundle and stored in a plastic bag.

HAND NETS

An insect net with a long, extensible aluminium handle (no. 324 Tropics net, BioQuip Products, P.O. Box 681, Santa Monica, California, USA) is useful to collect bats in caves, mines, and buildings. This particular net has a 4 m (12 ft) handle composed of six pieces, each 60 cm (2 ft) in length, which screw together. An ordinary insect net can be modified for bat collecting by making a long handle of suitable material. You can also improvise with a coat hanger or a wire of similar gauge bent into a hoop and laced with mosquito netting or cheesecloth.

SHOOTING

Although less preferable to netting because of potential specimen damage, shooting is a useful technique for collecting some of the sac-winged (Emballonuridae), vespertilionid (Vespertilionidae), and free-tailed (Molossidae) bats that are difficult to net because they forage above the tree canopy and for bats that roost in large caves with high ceilings or in tall palm trees. For minimum damage to specimens, use fine shot (no. 12 shot) loaded in .22 or .32 calibre rifle shells or in .410 gauge shotgun shells. For best results .22 rifle shells loaded with no. 12 shot should be used in a special smooth bore .22 gun. Auxiliary barrels ("Aux.") that slip into 20, 16, or 12 gauge shotgun chambers are made for holding .32 gauge or .410 gauge shells loaded with no. 12 shot.

BAT TRAPS

Several designs for bat traps have been used but one of the most versatile is the Tuttle trap. This trap consists of two rectangular aluminium frames that support vertical wire strands. Bats collide with the wires and fall unharmed into a canvas collecting bag. The Tuttle trap is not available commercially; however, for a description, including specifications for construction, see Tuttle (1974). In arid regions with restricted water, a single strand of fine piano wire can be stretched across a pond or water tank about 20 cm (6 inches) above the water. Bats striking the wire will fall into the water. As bats swim to shore, they can be captured with a hand net.

B. Collecting Techniques

Bats are collected primarily from diurnal roosts and from foraging or drinking sites at night.

ROOSTS

Caves and Mines

Caves and mines (especially abandoned ones) are usually productive sites for the bat collector. Larger caves or mines that may contain bats may be located from large-scale topographic maps (1:50 000), which usually indicate cave or mine sites, and by questioning local residents. The collector may save considerable time and effort by hiring a local person familiar with a particular mine or cave to act as a guide.

The cave or mine should be entered slowly with a minimal amount of noise. Some species roost in dim areas near the entrance; others prefer dark areas deep within. Examine each hole or depression in the roof as well as side openings while listening for vocalizations or for sounds of flying bats. Bats roosting on the ceiling may be taken in a long-handled insect net or shot—if there appears to be no danger of the ceiling's collapsing. Bats flying about can be captured in bat traps or with small pieces (1 m; 2–4 ft) of old mist net strung across corridors. Mist nets set near the cave or mine entrance before dusk may capture large numbers of bats as they leave to forage.

Buildings

Many bats, particularly some vespertilionids (Vespertilionidae) and free-tailed bats (Molossidae) roost in tile, thatched, or metal roofs, attics, and cavities between walls of buildings. Local inhabitants can usually provide information on buildings that contain bat colonies. Bats roosting in buildings may often be captured by hand (use gloves), with hand nets, or with long forceps (25 cm; 10 inches) that will reach into holes and crevices. If you cannot capture bats from buildings during the day, it may be possible to collect them with mist nets, hand nets, or bat traps as they leave the building to forage at night. By carefully observing the building in the evening, you can usually locate the openings that bats are using for exits.

Other Roosts

During the day, some bats roost in hollow trees or logs, under the bark of trees, in rock crevices, under large leaves such as palm or banana fronds, in culverts, under bridges, and even under rocks or stones. Migrating tree bats and Old World fruit bats may hang from trees and bushes in open, exposed areas.

FORAGING SITES

Although foraging bats can be collected with traps or by shooting, generally the most effective technique is mist netting. To net foraging bats efficiently, the collector should become familiar with the most productive areas. Many species can be netted near their feeding sites (orchards, wild fruit trees, flowering shrubs, trees, and over ponds or streams). Forest trails, the edges of forested areas, and highland passes are often used by bats as natural flyways. Isolated ponds in arid regions may attract great numbers of bats, particularly during the dry season. Nets set over streams or forest trails should be positioned in narrow areas where natural obstacles funnel bats into the net. Factors that will reduce netting success are rain, heavy dew, and moonlight. Some experimenting with the height of the net above the ground, the angle of the net relative to a flyway and its position relative to surrounding vegetation is necessary to obtain satisfactory results in different localities. Many species fly just above the ground vegetation so that nets set with the bottom strand about 20 cm (6 inches) above the ground is a good starting point from which to experiment.

3.2 Other Small Mammals

Snap Traps

The most successful trap for catching small rodents, marsupials, and insectivores is the snap trap. These traps are generally sold commercially (Victor Traps, Litiz, Pennsylvania, USA) in mouse-trap and rat-trap sizes. The larger, more powerful rat-trap is designed for killing mammals the size of rats and small squirrels. Designed for mammals of shrew and mouse size, the smaller mouse-trap is more effective, but the spring bar on the trap frequently crushes the skull of the specimen.

Most collectors prefer the Museum Special model (Fig. 2) (Woodstream Corporation, Litiz, Pennsylvania, USA). This trap is intermediate in size between the

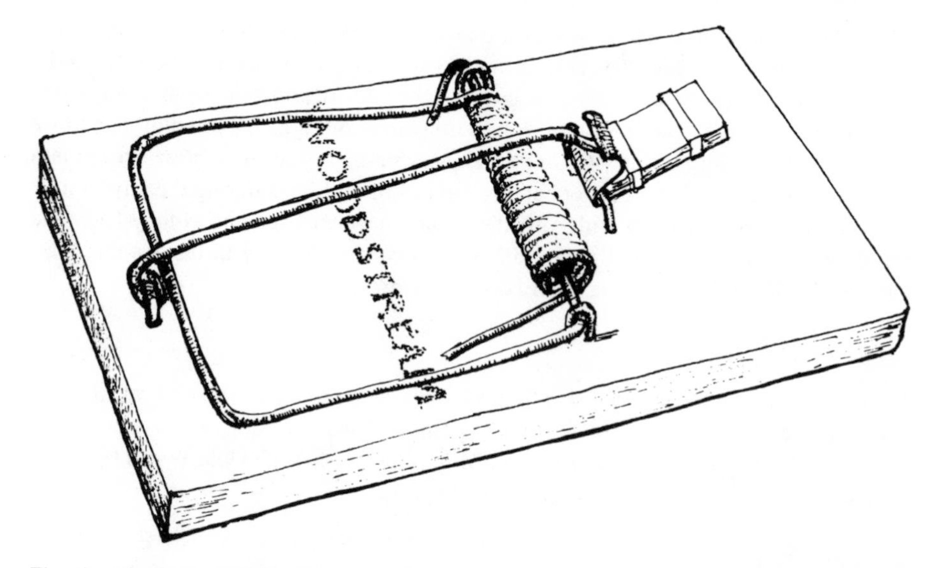

Fig. 2 Museum special snap trap shown in the set position. Trap size is 14 cm × 7 cm (5.5 inches × 2.7 inches).

mouse- and rat-trap and it is designed to kill mammals weighing up to 50 g (2 oz). Advantages of this trap are an extremely sensitive trigger mechanism and a spring bar designed to break the specimen's back rather than to crush the skull. Equipped with a weak spring, this trap can be used to catch small mammals such as shrews without seriously damaging the specimen. Snap traps should be checked at least once every 24 hours, preferably in the early morning. Dead mammals decompose rapidly, especially in warm weather. In tropical areas you may have to check traps more frequently as ants may quickly eat specimens.

Live Traps

Live traps are used to obtain live mammals for karyotyping, biochemical analyses, and parasite studies. Also, live traps permit the collector to select only those mammals required for specimens and release others unharmed. As some animals (e.g., shrews) may be reluctant to enter, live traps are generally not as effective as snap traps. To sample the mammalian fauna of an area accurately, you should supplement live trap lines with some snap traps.

For mammals such as squirrels, hares, and foxes welded wire mesh traps are available in several sizes (Havahart Company, P.O. Box 551, Ossining, New York, USA; National Trap Corporation, P.O. Box 302, Tomahawk, Wisconsin, USA). These traps usually have a front and rear door and are activated by a bait pan in the centre of the trap.

For small rodents and some insectivores the most popular live traps are the Longworth (Longworth Scientific Instrument Company Limited, Thames Street, Abingdon, Berkshire, England) and the Sherman trap (H. B. Sherman Company, Box 683, DeLand, Florida, USA). The aluminium Longworth trap has a trigger mechanism with a detachable nest box. The Sherman style trap, a rectangular box constructed from aluminium or galvanized metal, has a spring-loaded treadle which releases the door when depressed. An assortment of sizes and models, including folding and nonfolding are available. We have found a modification of the Sherman trap (Canadian Penitentiary Industries, Sir Wilfrid Laurier Building, 340 Laurier Avenue West, Ottawa, Ontario, Canada K1A 0P9) to be effective in catching mammals ranging in size from shrews to small squirrels. The trap is a nonfolding type with a screened rear door (Fig. 3).

To prevent deaths in live traps, check them at least once daily, preferably early in the morning. Also place some cotton wool for nesting material inside traps during cool weather.

Bait

An effective bait for small mammals that can be used in snap traps or live traps is a mixture of peanut butter and rolled oats. Chopped nuts, seeds, bits of chopped fruit (apples, raisins, bananas), or cheese can be added to this mixture. Experiment with different combinations of bait to determine the one most effective in a particular area. Plastic containers with screw-top lids or plastic squeeze tube containers are useful for carrying premixed bait when checking traps.

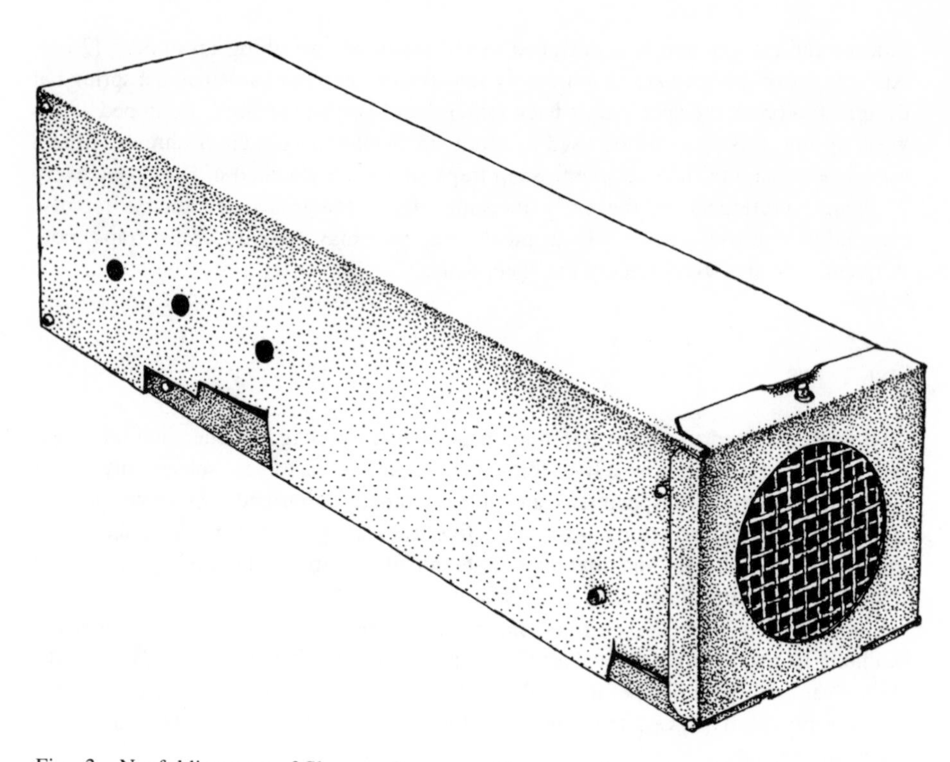

Fig. 3 Nonfolding type of Sherman live trap with screened rear door. Trap size is 30 cm × 8 cm × 8 cm (12 inches × 3 inches × 3 inches).

Special Traps

Special traps are made (Victor Traps, Litiz, Pennsylvania, USA; Z. A. MacAbee Gopher Trap Company, 110 Loma Alta Avenue, Los Gatos, California, USA) for capturing such fossorial mammals as pocket gophers (Geomyidae) and moles (Talpidae). Gopher traps (Fig. 4) are set in tunnels that are 20 to 900 cm (6–36 inches) below the surface of the ground. Locate the shallow tunnels near freshly discharged earth, remove a section of the tunnel, and set a pair of traps in each direction in the runway. When excavating earth to repair the opened tunnel, the pocket gopher will set off the trigger mechanism of the trap. To prevent traps from being dragged into the burrow system, tie them with wire to a firm stake.

Of the various mole traps sold commercially, the harpoon type (Fig. 5) appears to be the most efficient. Push the trap into the ground over the mole tunnel and flatten the raised tunnel so that the trigger pan will be set off by the mole as it moves through the tunnel. Consult Baker and Williams (1972) for a description of a live trap for gophers and Yates and Schmidly (1975) for a mole live trap.

A simple, effective trap for shrews and certain species of rodents (e.g., jumping mice and microtine rodents) is a fruit juice or similar-sized can 35 cm × 18 cm (14 inches × 7 inches) set flush with the ground. Cone-shaped pitfall traps, which have the advantage of being easy to carry in the field because they nest together, are sold commercially (Northwest Metal Products Company, P.O. Box 10, Kent, Washington, USA). It is not necessary to bait pitfall traps as the mammal simply

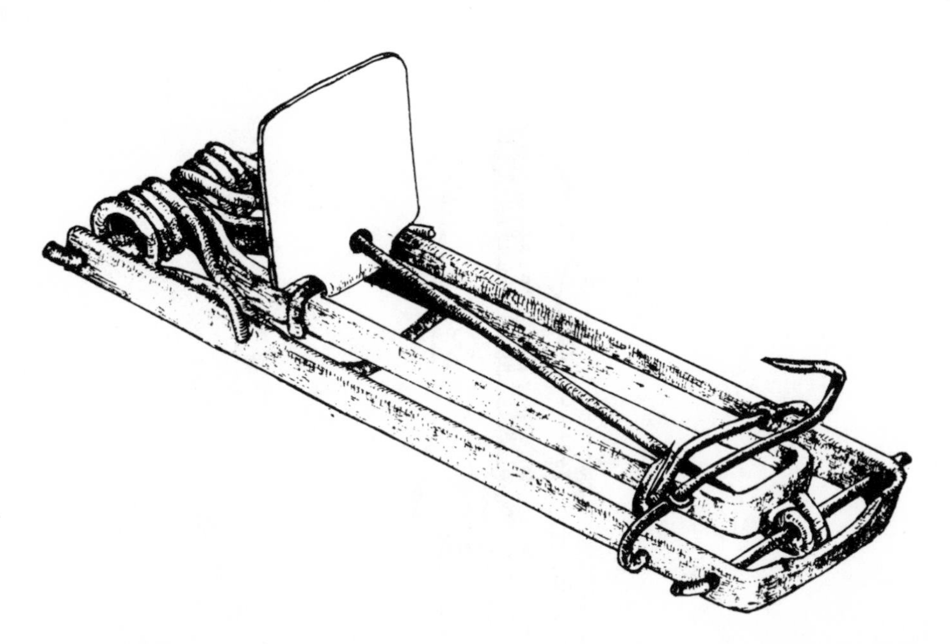

Fig. 4 Pocket gopher trap in the set position. Trap size is 15 cm × 5 cm (6 inches × 2 inches).

tumbles into them. About 75 mm (3 inches) of water in the trap will prevent rodents from jumping out. Set them adjacent to streams, in runways, or near burrows for best results. When sampling a habitat, set at least a few pitfall traps for they frequently capture species not taken in other types of traps.

Operating a Trap Line

To sample the small mammals of an area thoroughly, traps should be set in a variety of habitats (forest, open grassy area, transition zones). The recommended procedure is to set the traps in a "trap line" at regular intervals and roughly in a straight line. Trap sites are marked by tying a piece of coloured, plastic flagging tape or strip of cloth to a tree branch or a clump of vegetation. The total number of traps in a line (usually 30–100 traps), the number of traps at each site, and the spacing of traps is determined by experience. For most habitats, you will obtain good results by setting traps about 10 m (30 ft) apart with two or three traps at each site. To be certain that no trap sites are missed when checking a trap line, many collectors number their traps in sequence. Permanent numbers can be painted on live traps and numbers can be written with a pencil on wooden-based snap traps. Another method is to write trap site numbers on plastic flagging tape with a felt-tipped marker pen (waterproof ink type).

Rather than randomly selecting trapping sites, carefully choose the most favourable microhabitat, for example, the base of trees or stumps, on top of logs, in conspicuous runways, at burrow entrances, or at the edge of streams or ponds. Many small mammals confine their movements to runways that appear as well trampled miniature trails in the vegetation. Other signs indicating the presence of small mammals are droppings, tracks, piles of cut grass or sedges, and seed caches. In tropical forests

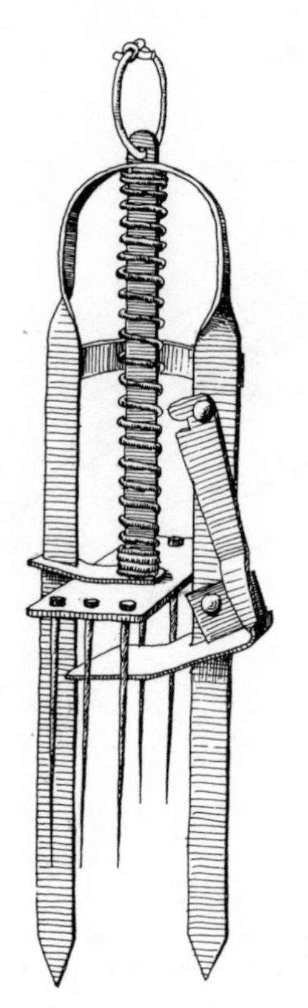

Fig. 5 Harpoon type of mole trap. Trap length is 40 cm (16 inches).

some species of small mammals are arboreal and traps set in the branches of trees will catch species not otherwise taken.

SHOOTING

For mammals such as rabbits, squirrels, and small- to medium-sized carnivores, shooting may be more effective and humane than trapping. To prevent excessive damage, use a small gauge shotgun (.410 or 20 gauge) loaded with light shot instead of a .22 rifle. BB shot or no. 2 shot is recommended for fox-sized animals, shot no. 4, 6 or 7½ is suitable for hare-sized, and no. 11 or 12 shot is used for smaller mammals (pikas, squirrels). Auxiliary barrels are discussed in section 3.1

3.3 Fur-bearers and Large Mammals

Steel leg-hold traps, Conibear traps, snares, and shooting are the usual methods for collecting these mammals but check local laws and regulations before doing so. Arrangements with experienced fur trappers to secure specimens taken in season often provide good results. For reviews of the techniques used for trapping fur-bearers, consult *The Manitoba Trappers' Guide* (Manitoba Department of Renewable Resources and Transportation Service 1965) or Stains (1962).

4. *Documenting Specimens*

Described in this section are basic field data that should be routinely recorded by *anyone* collecting mammals for scientific specimens (field number, date collected, nature of specimens, measurements, sex, reproductive data, locality descriptions, habitat descriptions, method of capture, and miscellaneous field notes). However, it cannot be over-emphasized that this information must be accurate. If you are not certain, *do not guess!* If you are uncertain of the sex, for example, then indicate so with a question mark on your catalogue or field notes.

4.1 Recording Data

Collectors usually record their field data in a notebook. DeBlase and Martin (1974) and Hall (1962) recommended that field notes be organized into three sections: (1) a journal; (2) a catalogue; (3) species accounts. At the ROM we use a slightly different system. Specimen data are recorded on printed catalogue sheets (Fig. 6). Similar or modified sheets can be designed to suit your particular needs. The following is a brief explanation of headings of columns shown on the catalogue sheet in Figure 6.

The museum number is assigned to the specimen when it is accessioned by the museum, consequently leave this space blank. Each specimen listed in the catalogue must have a separate field number and this number is also written on a tag (in pencil or waterproof ink) that is securely tied to the corresponding specimen (see section 4.2). A tentative identification, even a common name of the specimen, should be entered under the species heading. Always write the date with the month in full, that is, June 10, 1965 *not* 10/6/65. Documenting the categories for sex, measurements, and locality are discussed in sections 4.4, 4.3, and 4.6. The nature of the specimen (skin and skull, skull only, skin only, skeleton only or preserved in formalin), should also be listed in the remarks for *each* specimen.

Such information as reproductive data, habitat, and field observations (see sections 4.5, 4.7, and 4.9) may be entered in the remarks section; however, usually it is impossible to fit all of this information on the catalogue sheet. We recommend that a field notebook or diary be kept in conjunction with the catalogue sheets. To ensure that data are associated with the appropriate specimens, carefully list the field numbers with their corresponding data. Figure 7 illustrates a page taken from a typical field notebook. At the ROM we also use printed sheets for recording reproductive data (Fig. 8). You may wish to design similar sheets or simply record these data in your notebook.

All catalogues, field notes, photographs, and maps of collecting sites are kept permanently by museums as documentation for researchers. As catalogue sheets and field notes are often the only source of information for specimens, your catalogue and notes should be well organized, legible, and as accurate as possible. Write in pencil or waterproof ink as other types of ink will run if subjected to moisture. Catalogue sheets or notes that have been seriously damaged from moisture, grease, or blood should be recopied.

1977 MAMMAL CATALOGUE ROYAL ONTARIO MUSEUM

Year

Collectors: Nagorsen + Eger Page: 2

General Area: Fort Frances District, Ontario, Canada

Museum Number	Field No.	Species	Sex	TL	TV	HF	Ear	FA	WS	WT	Locality	Date	Remarks
	8591	Peromyscus maniculatus	A ♂	171	85	21	20	—	—	19.5	Long Sault Rapids, Rainy River, Roseberry	Aug. 2	Study skin + skull snap trapped
	8596	"	A ♀	173	88	20	20	–	–	18.5	Twp. 48°38'N 94°06' W	"	flat skin + skeleton snap trapped
	8597	"	S ♀	158	80	20	20	–	–	14.0	"	"	preserved in formalin live trapped; subadult
	8598	Clethrionomys gapperi	A ♀	124	39	18	15	–	–	15.5	"	"	Snap trapped skeleton only
	8600	"	A ♂	137	43	18	15	–	–	23.0	"	"	snap trapped; mature study skin + skull
	8601	"	J ♂	110	30	17	13	–	–	11.0	"	"	live trapped; immature preserved in formalin
	8602	Sorex cinereus ?	A ♂	96	41	11	6	—	-	5.0	"	"	snap trapped study skin + skull
	8603	Sorex arcticus	A ♀	105	41	14	7	-	-	8.2	"	"	snap trapped skeleton only
	8604	Tamias striatus	A ♂	251	96	36	17			75.0	"	"	live trapped, Study skin + skull
													2 botfly larvae from inguinal region preserved.
	8605	Lepus americanus	A ♀	432	-	141	71	–	–	kg 1.6	2 mi. E. of Gameland, Pratt Twp. 48°52'N, 94°27'W	Aug. 2	Shot; flat skin + skeleton. ticks from head preserv.
	8606	Myotis lucifugus	A ♂	85	32	9	13/7	38		7.0	Little Grassy River, Bergland, McCrossan	"	netted 2100 hrs preserved in formalin
	8607	Eptesicus fuscus	A ♀	101	40	12	17/	43		14.0	Twp., 48°57'N, 94°23' W.	"	netted 2100hrs preserved in formalin

Fig. 6 Example of a catalogue sheet used by ROM collectors for recording specimens and their corresponding field numbers, measurements, localities, and dates of collection.

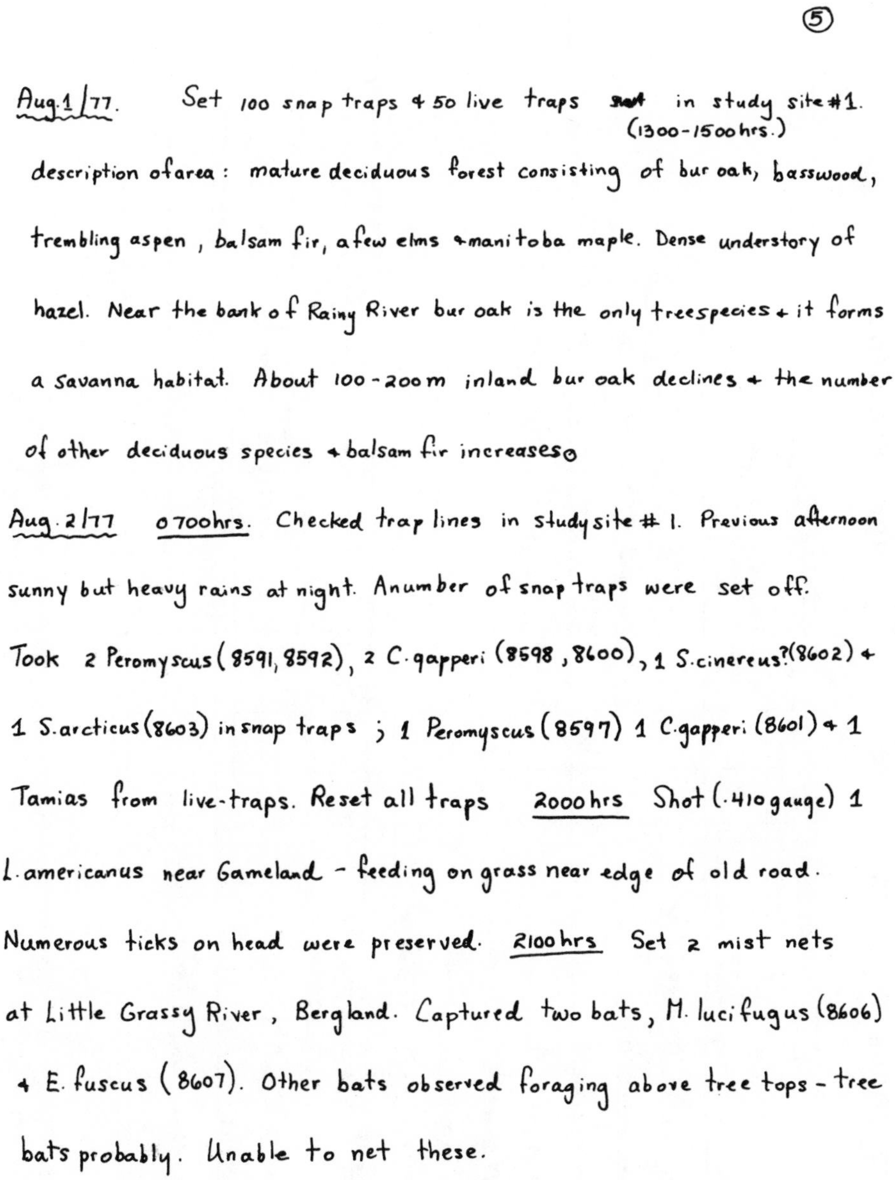

⑤

Aug.1/77. Set 100 snap traps + 50 live traps in study site #1. (1300-1500 hrs.)

description of area: mature deciduous forest consisting of bur oak, basswood, trembling aspen, balsam fir, a few elms + manitoba maple. Dense understory of hazel. Near the bank of Rainy River bur oak is the only treespecies + it forms a savanna habitat. About 100-200 m inland bur oak declines + the number of other deciduous species + balsam fir increases.

Aug. 2/77 0700hrs. Checked trap lines in study site # 1. Previous afternoon sunny but heavy rains at night. A number of snap traps were set off. Took 2 Peromyscus (8591, 8592), 2 C. gapperi (8598, 8600), 1 S. cinereus? (8602) + 1 S. arcticus (8603) in snap traps; 1 Peromyscus (8597) 1 C. gapperi (8601) + 1 Tamias from live-traps. Reset all traps 2000 hrs Shot (.410 gauge) 1 L. americanus near Gameland - feeding on grass near edge of old road. Numerous ticks on head were preserved. 2100 hrs Set 2 mist nets at Little Grassy River, Bergland. Captured two bats, M. lucifugus (8606) + E. fuscus (8607). Other bats observed foraging above tree tops - tree bats probably. Unable to net these.

Fig. 7 A page taken from a field notebook of a ROM collector to illustrate the data recorded. Specimens are the same as those listed in the catalogue sheet (Fig. 6).

4.2 Specimen Tags

All specimens listed in your catalogue should be labelled with their corresponding field numbers. DeBlase and Martin (1974) described some of the different kinds of tags used by field collectors. ROM collectors use printed field number tags (Fig. 9D) that are resistant to alcohol and formalin to label study skins, skulls, skeletons, or

REPRODUCTIVE DATA — ROM SPECIMENS

			MALES			FEMALES							
				Epididymes		Vagina							
Field Number	Species	Age	Testes Size	Visible	Not Visible	Perf	Imp	Normal Embryos #	Resorbed Embryos #	CR of Normal Embryos	Placental Scars #	Mammary Tissue	Remarks
8591	Peromyscus maniculatus	A	12x9mm	✓									
8596	Peromyscus maniculatus	A				✓		2L,3R	1L (CR=3mm)	10mm	3L,2R	heavy with milk	
8597	Peromyscus maniculatus	S					✓	—	—	—	—	—	nulliparous
8598	Clethrionomys gapperi	A				✓		2L,4R	0	6mm	0	none	
8600	Clethrionomys gapperi	A	13x8mm	✓									
8601	Clethrionomys gapperi	J	3x2mm		✓								
8602	Sorex cinereus	A	8x3mm	✓									
8603	Sorex arcticus	A											organs decomposed
8604	Tamias striatus	A	8x4mm		✓								
8605	Lepus americanus	A				—		3L,3R	0	35mm	present *	heavy with milk	* two sets, can't count
8607	Eptesicus fuscus	A											nulliparous
8606	Myotis lucifugus	A	3x1 mm		✓								

Fig. 8 Example of a data sheet used by ROM collectors for recording reproductive data. Specimens are the same as those listed on the catalogue sheet (Fig. 6) and the field notes (Fig. 7).

specimens in fluid. If these are not available, you can construct similar tags from vinyl-coated paper or white cardboard (Fig. 9C). When labelling fluid-preserved specimens, ensure that tags are resistant to preserving solutions. A recommended method for coding field numbers is to write your initials preceding the specimen number (e.g., JHW 237). This practice will prevent any possible confusion with specimens of another collector.

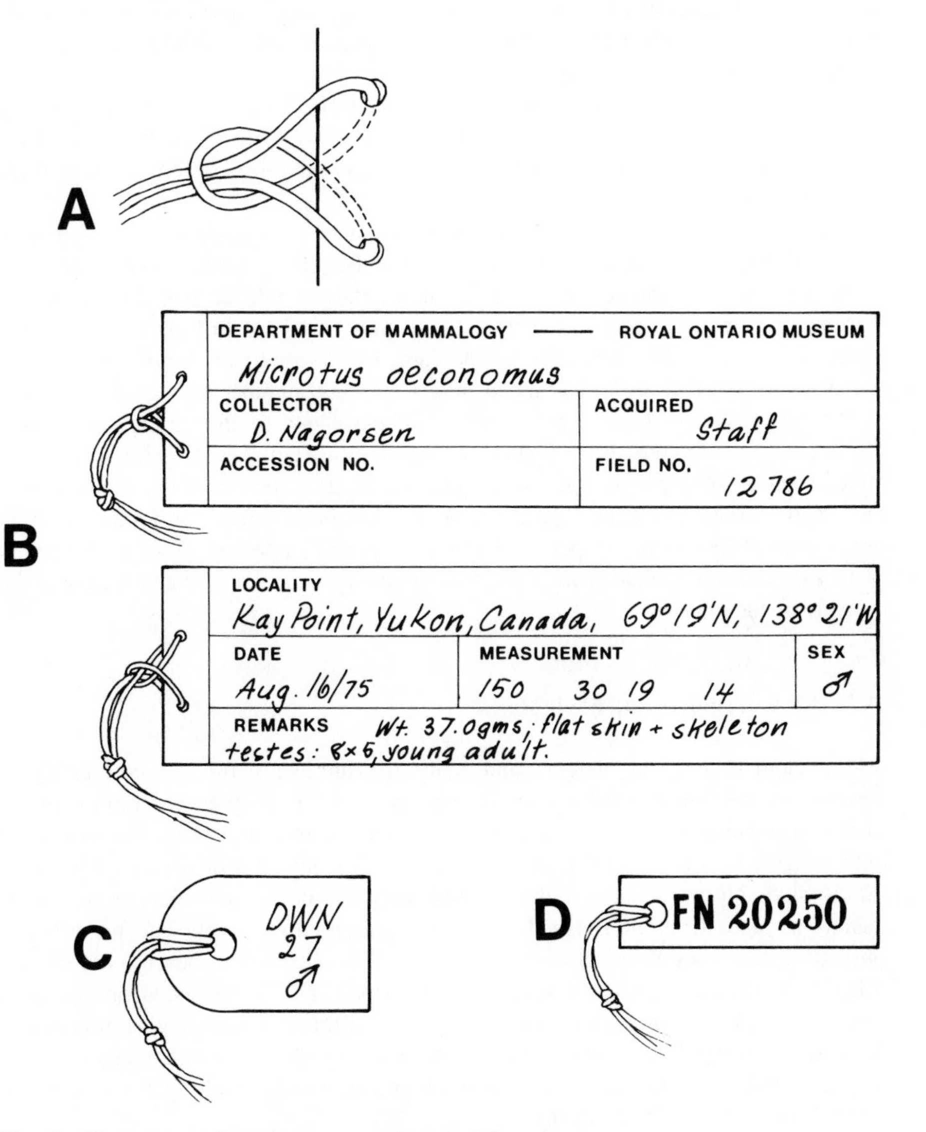

Fig. 9 Three types of specimen tags and method for tying.

A Knot used for stringing study skin tag.

B Study skin label used by many field collectors. Information recorded includes locality, date, sex, collector's name, measurements, reproductive data, identification of specimen, and field number.

C Label constructed from vinyl-coated paper; field number consists of collector's initials and specimen number.

D Paper label (alcohol and formalin resistant) with printed field number.

For study skins and entire specimens in fluid, some collectors use a label similar to that illustrated in Figure 9B. Information recorded on these labels includes locality, date, sex, collector's name, identification of specimen, measurements, and reproductive data. Standard measurements (see section 4.3) are always listed in the following sequence: (1) total length; (2) tail vertebrae length; (3) hind foot length; (4) ear length; and in bats (5) tragus length, and (6) forearm length. At the ROM these labels are now generated by our computerized cataloguing system. As a result, we do not fill out these labels in the field and study skins and fluid-preserved specimens are labelled with field number tags (Fig. 9D).

To prepare the tag for tying, run a thread through two holes near one end of the tag and tie a knot about 2.5 cm (1 inch) from the tag (Fig. 9A). You can save considerable time in the field by stringing your tags with thread before you take them into the field.

Tie tags on study skins and specimens in fluid above the ankle with a square knot (Fig. 37). For flat skins, also write the field number on the cardboard stretcher with waterproof ink. If printed field tags are used, they should be tied directly to the cardboard (Fig. 41). Some collectors (Anderson, 1965) record such data as sex, measurements, locality, and date on the card. For "cased" or "open" skins from larger mammals that are to be tanned, attach the field number through the nostrils.

If skulls or skeletons are stored in individual gauze bags for drying, tie a tag around the neck of the bag (Fig. 39B). If skulls or skeletons are not placed in individual bags, attach the tag directly to the specimen. Skull labels can be tied around the mandible or in larger mammals to the zygomatic arch. A suitable area for tying labels to articulated skeletons is the pelvis. For disarticulated skeletons, label each portion with a tag. If large skeletons are placed in burlap bags, you can also attach a tag to the outside of the bag.

4.3 Measurements and Weights

Mammalogists rely on weights and body measurements for aid in identifying specimens, determining the age of specimens, and for studying variation between different populations. It is essential that the collector record accurate measurements and weights *before* the specimen is prepared as a study skin or preserved in fluid. Study skins shrink somewhat during their preparation and reliable measurements cannot be made from the finished skin. Fluid-preserved specimens become stiff and inflexible once they have set in the fixative and are difficult to measure accurately. Measurements used in research papers and descriptions of species are always given in metric units. Linear measurements should be in millimetres and weights in grams or kilograms. Indicate an approximate measurement by the *circa* abbreviation "ca", e.g., ca 180 mm. Also note any aberrant measurements resulting from damaged specimens (tail broken, ear torn).

MEASUREMENTS

The following are the standard measurements taken by North American collectors (Fig. 10). European collectors frequently record head and body length (HB) rather than total length and omit the claw for the hind foot measurement. For the sake of

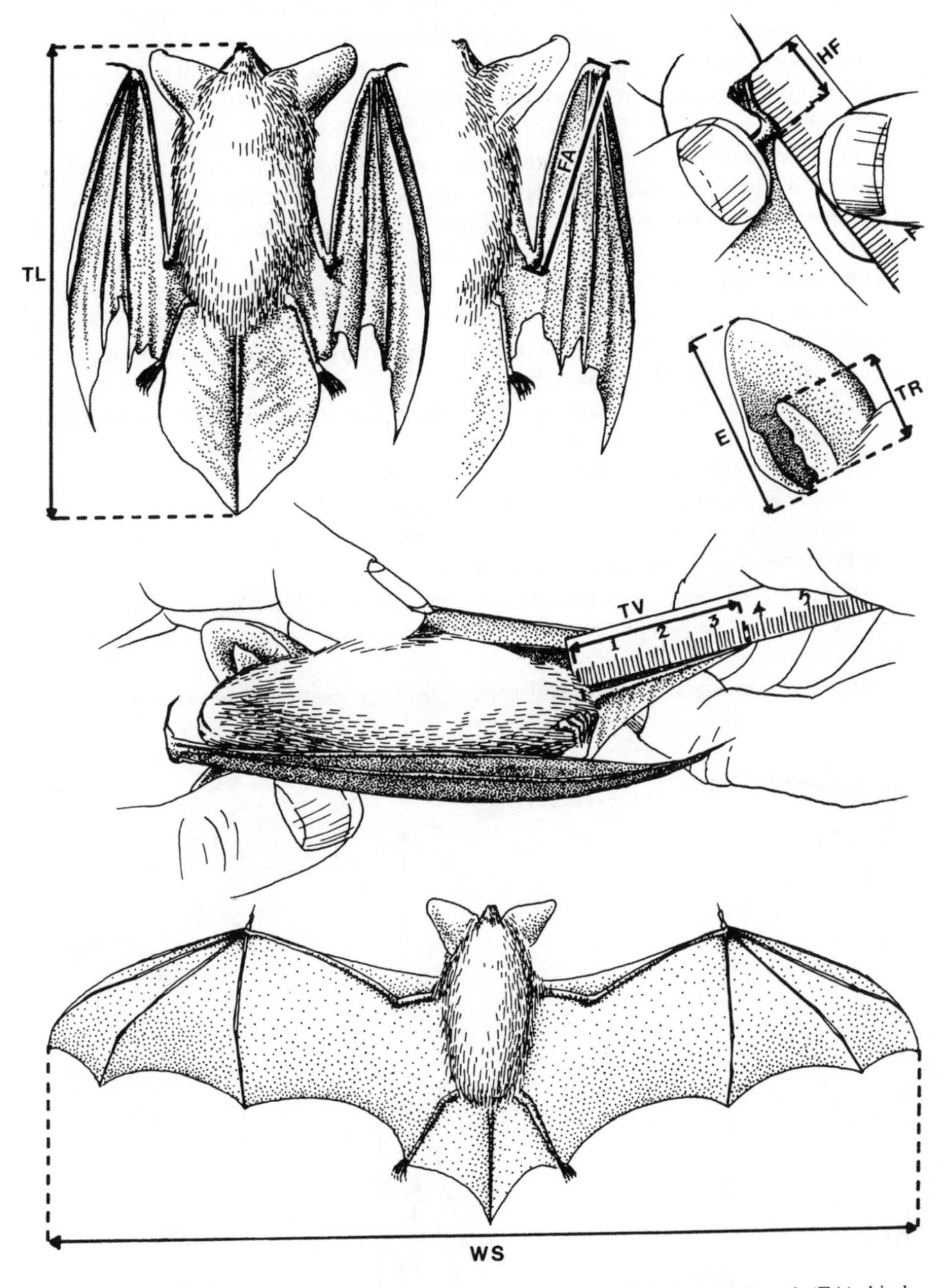

Fig. 10 Standard body measurements for bats. Total length (TL), forearm length (FA), hind foot length (HF), tragus length (TR), ear length (E), tail vertebrae length (TV), and wingspan (WS). For other small mammals, only the total length, tail vertebrae, hind foot, and ear are recorded.

consistency, we recommend that collectors use the North American measurements.

Total Length (TL): straight-line distance from the tip of the nose to the end of the last tail vertebra, exclusive of the hairs. Lay the mammal on its back on the ruler and measure by extending the specimen, pressing the body flat. Pull the tail (if present) to its full length, measuring to the end of the last bone. If no tail is present, measure to the end of the backbone.

Tail Vertebrae (TV): distance from the base of the tail to the tip of the last vertebra, exclusive of the hairs. With the mammal on its belly, place the ruler at the point where the tail joins the body, pull the tail upward and measure to the end of the last vertebra.

Hind Foot (HF): distance from the end of the heel bone (calcaneum) to the end of the claw on the longest toe. Stretch the toes and measure from the heel to the longest length of the claws. For hoofed mammals, the HF measurement is taken from the tip of the hock to the tip of the hoof (Fig. 11).

Ear Length (E): distance from the base of the notch of the lower part of the ear to the uppermost margin of the ear.

Measurements limited to bats are:

Tragus (TR): distance from where the tragus joins the ear to its tip.

Forearm (FA): distance from the outside of the wrist to the outside of the elbow. Fold the wing when taking this measurement.

Wingspan (WS): distance between the wing tips when the wing is stretched out. Lay the bat on its belly and gently stretch the wings, being careful not to overstretch them.

Measure large mammals on level ground using a steel tape measure. The same

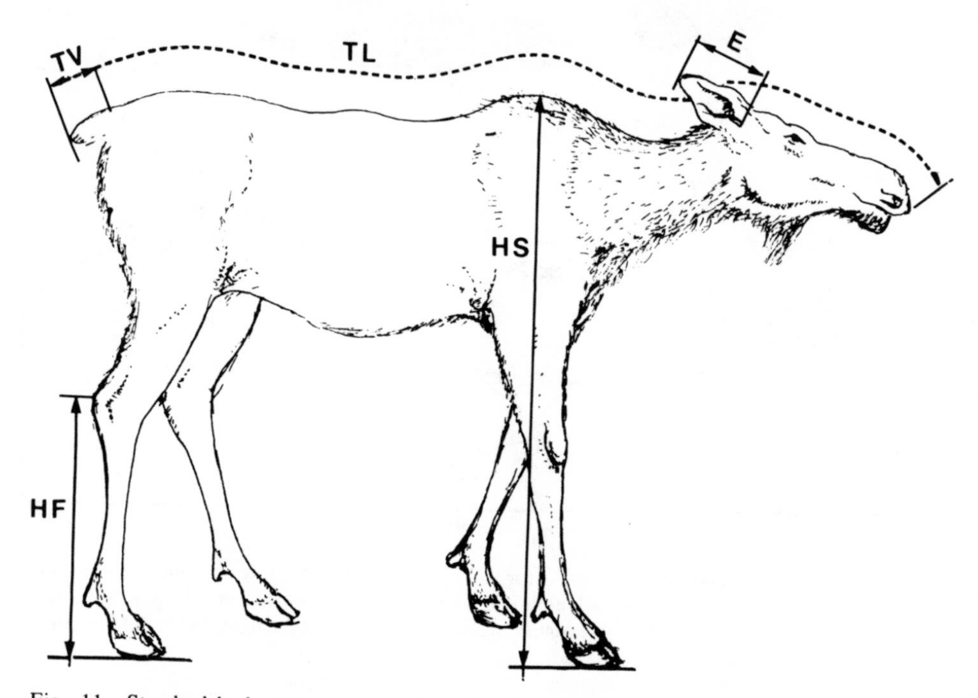

Fig. 11 Standard body measurements for a large ungulate. Total length (TL), tail length (TV), hind foot (HF), ear from notch (E), and height at shoulder (HS).

standard measurements (TL, TV, HF, E) are taken for these mammals (Fig. 11). An additional measurement is the height of the shoulder (HS) which is the distance from the top of the shoulder to the bottom of the foot. For total length, measure in a straight line with the body stretched out rather than measuring around the curves of the neck and back.

The measurements for cetaceans recommended by the American Society of Mammalogists are shown in Figure 12.

WEIGHTS

Specimens should be weighed promptly and before preparing. Weights should be taken in grams or kilograms; however, body weights in pounds or ounces are helpful when metric scales are not available. For approximate weights, use the abbreviation "ca". The body weight of a large mammal is difficult to obtain in the field;

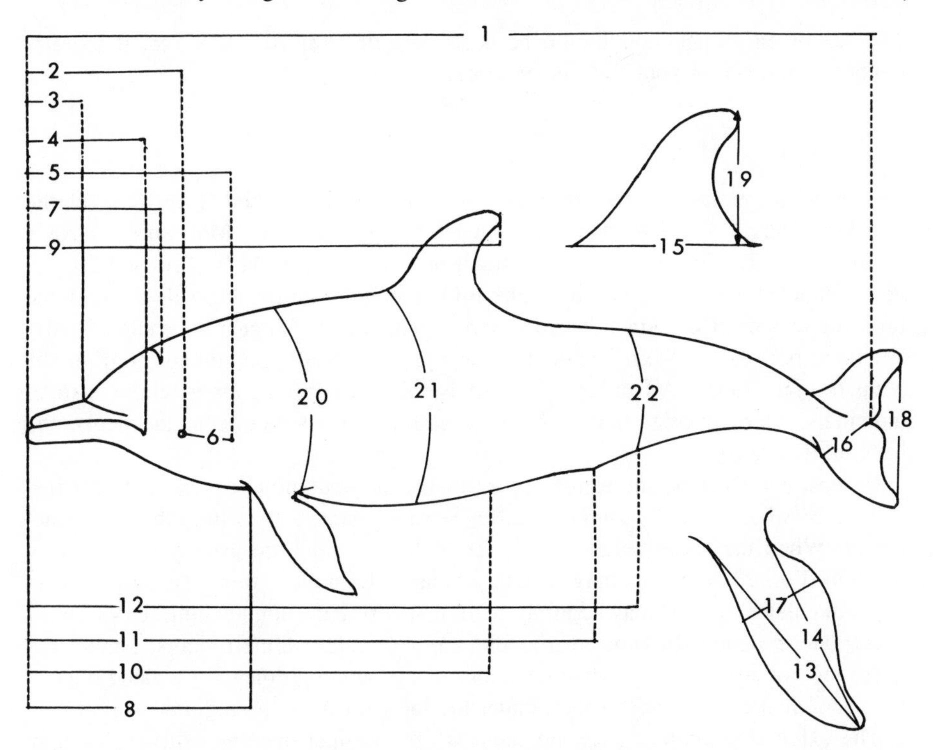

Fig. 12 Standard cetacean measurements as recommended by the American Society of Mammalogists. LENGTH: 1 total, 2 tip of upper jaw to centre of eye, 3 tip of upper jaw to apex of melon boss, 4 gape, 5 tip of upper jaw to external auditory meatus, 6 centre of eye to external auditory meatus, 7 tip of upper jaw to blowhole along midline or to midlength of two blowholes, 8 tip of upper jaw to anterior insertion of flipper, 9 tip of upper jaw to tip of dorsal fin, 10 tip of upper jaw to midpoint of umbilicus, 11 tip of upper jaw to midpoint of genital aperture, 12 tip of upper jaw to centre of anus, 13 anterior insertion of flipper to tip, 14 axilla to tip of flipper, 15 dorsal fin base, 16 distance from nearest point on anterior border of flukes to notch. WIDTH: 17 flipper (maximum), 18 flukes (tip to tip). HEIGHT: 19 dorsal fin (fin tip to base). GIRTH: 20 on a transverse plane intersecting axilla, 21 maxima, 22 on a transverse plane intersecting the anus.

nevertheless, collectors should weigh large mammals whenever possible because of the paucity of weight data.

High quality spring balances graduated in grams, for example, Pesola scales (Bleitz Wildlife Foundation, 5334 Hollywood Boulevard, Hollywood, California, USA) are excellent for the field collector. They are made in the following capacities: 5 g, 10 g, 30 g, 100 g, 500 g, 1000 g, and 2500 g. An Ohaus triple beam balance (Fisher Scientific Company, 711 Forbes Avenue, Pittsburgh, Pennsylvania, USA) may also be used for weighing mammals to 2000 g. Although more accurate than Pesola balances, the triple beam balance is heavier and less compact. For mammals greater than 2500 g use a heavy-duty spring balance (Forestry Supplier Incorporated, Cox 8397/205 West Rankin Street, Jackson, Mississippi, USA).

4.4 Determining the Sex of Mammals

The sex of each specimen should be accurately determined. However, if there is doubt, so indicate in your field notes or catalogue.

External Genitalia

The external genitalia of the male usually can be distinguished from those of the female by larger size and their position relative to the anus. Most males have a prominent penis (Fig. 13), but some small mammals, particularly shrews, have the penis retracted into a sheath of a tubular fold of skin during the intervals between the breeding seasons. Consequently, the external genitalia may appear to be superficially similar in both sexes. With fine pointed forceps, it is usually possible to protrude the penis from its sheath. A hand lens is often useful for examining the genitalia in small mammals. Many species have a bony or cartilaginous structure in the penis, the baculum (*os penis*).

In most adult males, the testes occur outside the abdominal cavity, but in a few mammals (whales and dolphins) the testes remain permanently within the abdominal cavity. When the testes occur outside the abdomen, they are usually situated in a scrotum (Fig. 13A). Testes may remain permanently in the scrotum (most primates, dogs, ungulates) or they may be intra-abdominal during the nonbreeding season (most bats, some rodents). In some mammals (shrews, moles, some rodents, hares), the testes are not contained in a distinct scrotum and, although they are located outside the abdominal cavity, they remain under the integument in the inguinal region.

The external genitalia of females consist of a vaginal opening (vulva) that may have prominent skin folds in some species. The urethra may also be visible (Fig. 13 C,D). During the breeding season, teats or nipples of the female mammary glands may be enlarged. The number and position of the mammary glands vary greatly in different species. In some mammals (primates and bats) there is only a single pair of mammary glands confined to the chest region. However, in mammals with numerous glands (rodents) the teats are usually situated in parallel rows along the ventral surface of the chest and abdomen (Fig. 13E).

In cetaceans (whales, dolphins, and porpoises), the genitalia are contained in a genital groove (Fig. 14). Normally the penis is retracted fully into a pouch in the ventral abdominal wall and only the genital slit can be seen. The penis protrudes from

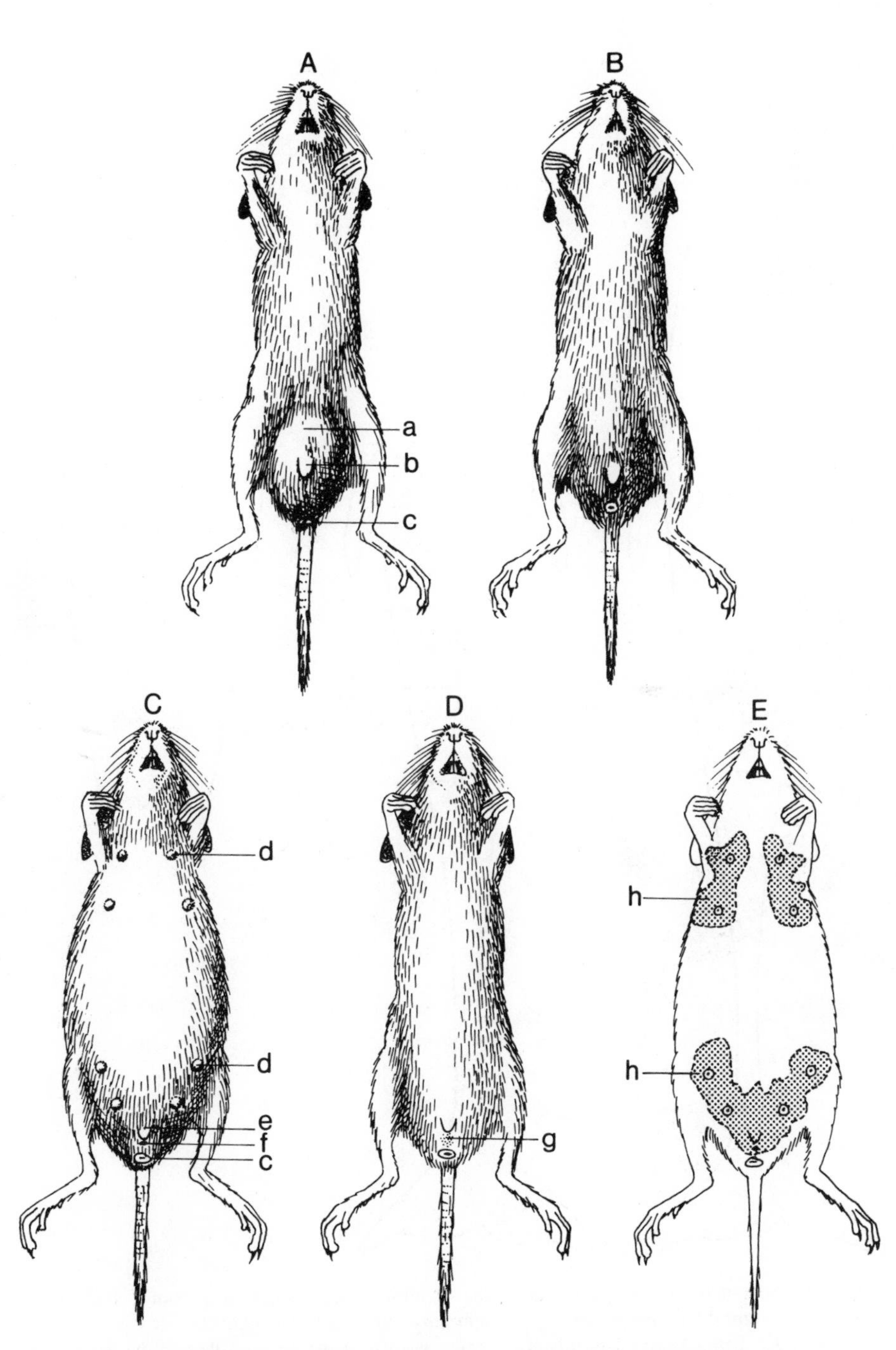

Fig. 13 External genitalia of a male and female cricetid rodent: a scrotum, b penis, c anus, d teats, e vulva, f vaginal opening (perforate), g vaginal opening (imperforate), h mammary tissue (stippled area).

A An adult male with an enlarged scrotum that partially obscures the anus.
B An immature male without an enlarged scrotum.
C Vulva, anus, and a perforate vagina on a pregnant female.
D Schematic drawing of a female showing a membrane covering the vaginal opening (imperforate condition).
E Schematic drawing showing the position of mammary tissue under the teats.

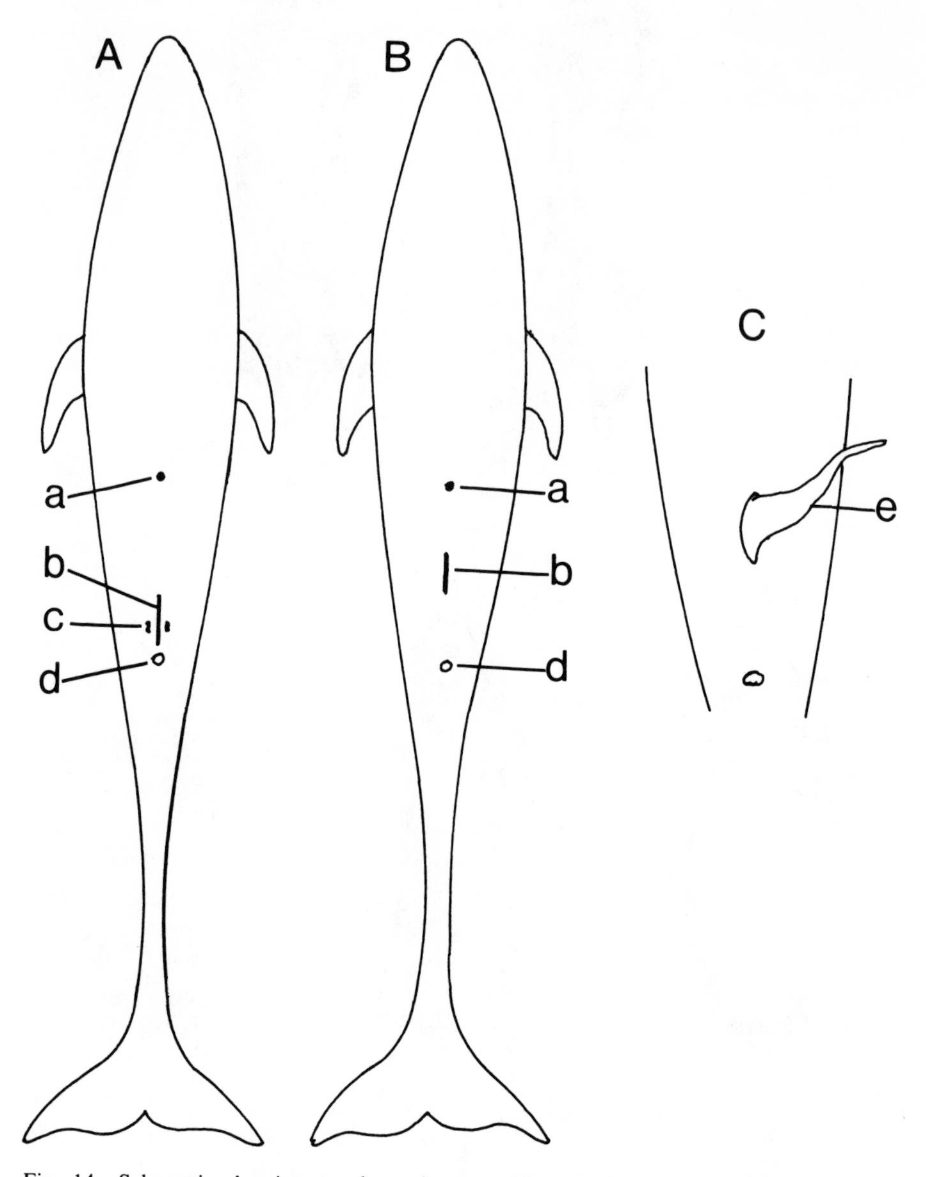

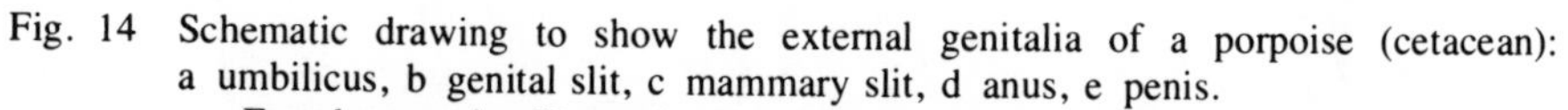

Fig. 14 Schematic drawing to show the external genitalia of a porpoise (cetacean): a umbilicus, b genital slit, c mammary slit, d anus, e penis.

A Female porpoise illustrating the presence of mammary slits and the position of the genital slit.

B Male porpoise with penis retracted into abdomen. Note position of the genital slit relative to the anus.

C Male porpoise with penis protruding from the genital slit.

the genital slit only during erection or occasionally on death. The female genitalia are also contained in a genital slit but the distance from the centre of the genital slit to the anus is much less in females than in males. In some porpoises, the anal and genital openings may occupy the same aperture. A number of accessory grooves may flank the female genital slit, but one pair—the mammary slits or grooves which contain the nipples—is always present and in a constant position in all species of cetaceans. The mammary slits are not present in male cetaceans.

Internal Reproductive Organs

After the skin has been removed, specimens to be prepared as study skins or skeletons should be dissected to verify the sex and to obtain reproductive data (see section 4.5). Testes appear as whitish or yellowish oval organs (Fig. 15). During intervals between the breeding season, testes may be small, especially in small mammals (shrews). Females can be distinguished by the presence of a uterus (Fig. 16).

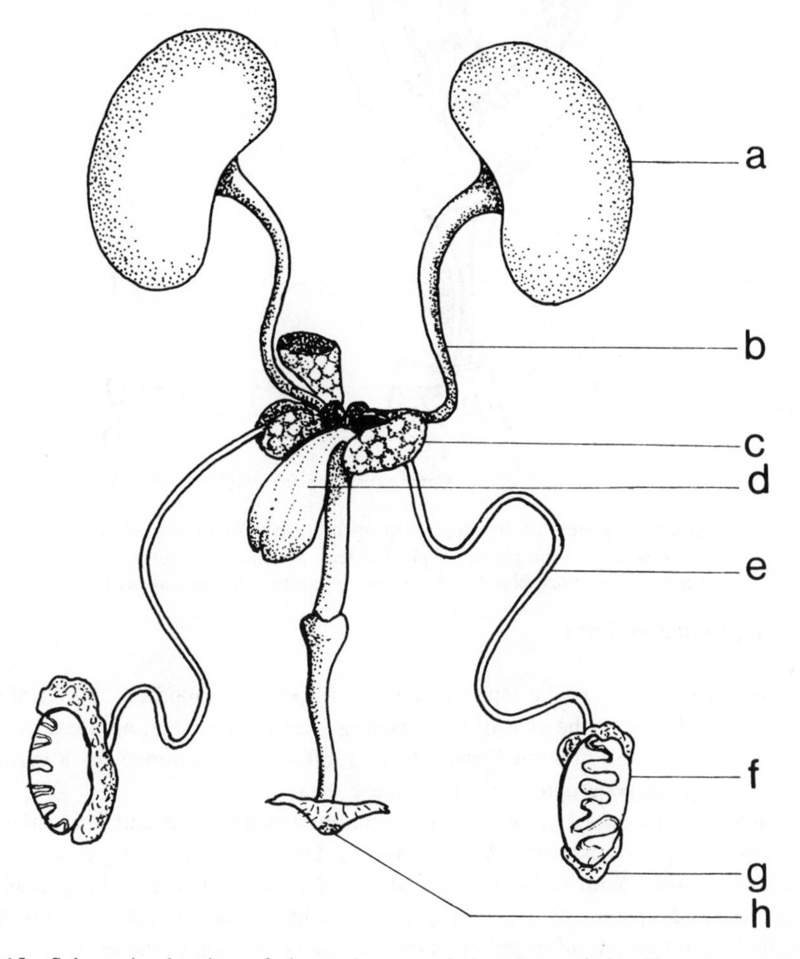

Fig. 15 Schematic drawing of the male urogenital system of the Norway rat (*Rattus norvegicus*) as an example of a typical rodent: a kidney, b ureter, c prostate gland, d urinary bladder, e vas deferens, f testis, g cauda epididymis, h penis.

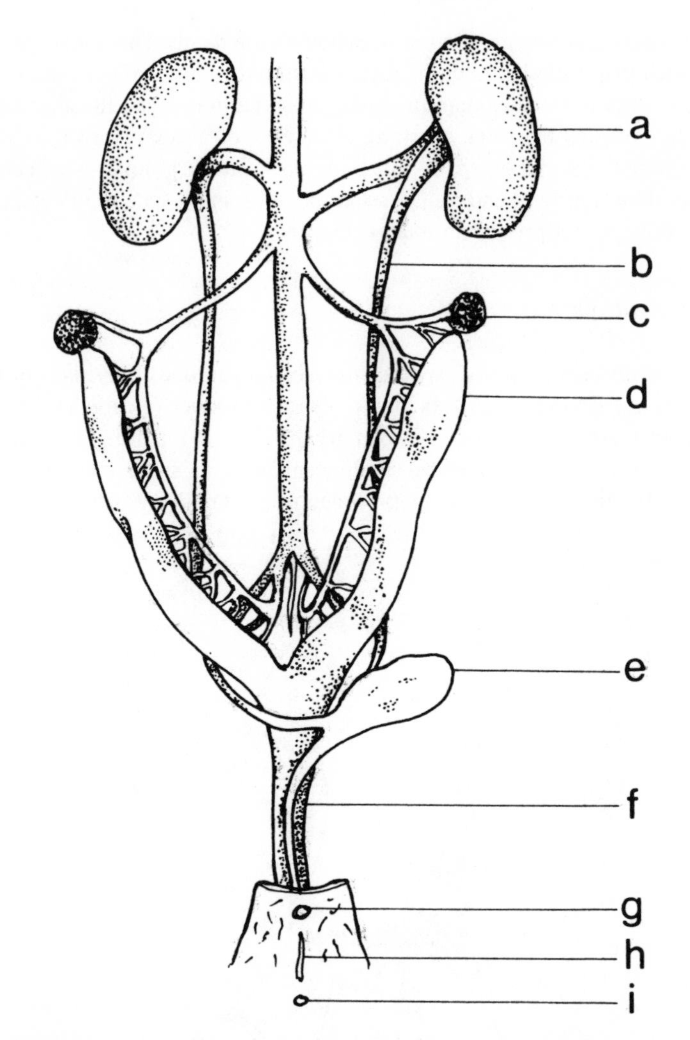

Fig. 16 Schematic drawing of the female urogenital system of the Norway rat (*Rattus norvegicus*) as an example of a typical rodent: a kidney, b ureter, c ovary, d uterus (left horn), e urinary bladder, f vagina, g urethra, h vaginal orifice, i anus.

4.5 Reproductive Data

Notes on the condition of the reproductive organs provide important biological data. The length and time of the year of the breeding season, litter size, numbers of litters per year, and the age of sexual maturity may often be determined for a particular species in a given geographic area from these data.

Mammals prepared as study skins or skeletons (sections 5.2 and 5.4) should be dissected immediately after the skin is removed. Open the body cavity by cutting the abdominal muscles with scissors or a scalpel and examine the reproductive organs. Fluid-preserved specimens (section 5.1) are usually not dissected in the field; nevertheless, important information on the breeding condition of the specimen can be provided by examining the external genitalia. If the condition of the reproductive organs cannot be determined because of decomposition, then indicate so in your field notes. *Do not guess* at the reproductive status! If you have difficulty distinguishing embryos or placental scars (see page 32), then preserve the uterus in a vial of 10 per cent buffered neutral formalin (with corresponding field number) and include it with your collection of specimens.

A. Males

Criteria used to distinguish breeding males are the size of the testes and the size of the tubules in the cauda epididymis. For dissected specimens, measure the length and width of the testes (both if different in size) in millimetres (e.g., 10 mm × 6 mm). For fluid-preserved specimens that have a scrotum (section 4.4), determine whether or not the testes are grossly enlarged.

Another useful criteria for breeding males in various small species (shrews, rodents, bats) is the occurrence of visible tubules in the cauda epididymis (Fig. 15). If the tubules are visible to the eye, they are swollen and usually contain sperm. However, if the tubules are not visible, they are probably void of sperm. For dissected specimens, note whether or not the tubules are visible and record the condition in your field notes.

B. Females

Breeding females may be pregnant, lactating, or both. Criteria used to diagnose the breeding condition of females are the condition of the vagina, presence or absence of embryos, presence or absence of placental scars, and the condition of mammary tissue.

CONDITION OF THE VAGINA

The vagina of females is usually sealed by a membrane until puberty. In some rodents and moles (Talpidae) that breed seasonally, the vagina may be closed by a membrane during the nonbreeding (anoestrus) of the reproductive cycle (Fig. 13D). The vagina in this condition is described as imperforate. In contrast, during heat (oestrus), the vagina is not sealed by a membrane and is perforate (Fig. 13C). In some mammals, for example the squirrels (Sciuridae), the vaginal region (vulva) may appear to be swollen or turgid during oestrus. The condition of the vagina is usually a reliable indication of reproductive condition, so note whether the vagina is imperforate or perforate and also whether the vulva is obviously swollen.

LACTATION

Lactation is defined as the secretion of milk. The following criteria are used as evidence that a female is lactating: (1) female is observed nursing young; (2) milk can be squeezed from teats; (3) heavy deposits of mammary tissue that contain milk are present. The criterion of milk in the mammary tissue can be only applied to dissected specimens. This mammary tissue is found on the inside of the skin in areas under the teats (Fig. 13E) and is usually whitish in colour and may spread beneath the skin some distance from the teats. Indicate in your field notes the presence of the mammary tissue and whether the tissue contains milk.

PREGNANCY

Pregnancy is defined as the condition of having a developing foetus or embryo. Females in late pregnancy may have a swollen abdomen and at this stage it may be possible to detect embryos by squeezing or pinching the abdomen.

In dissected mammals, carefully examine the uterus. When examining the uterus, it may be helpful to dissect it out and stretch it on a piece of white board or paper. For mammals with thick-walled uteri, you may have to open the uterus to examine it but this is seldom necessary for shrews, mice, or bats. The presence of embryos is positive evidence of pregnancy (Fig. 17C). Count and measure the crown-rump length (CR) of all embryos. This measurement (Fig. 17D) is from the top of the head to the end of the rump with the embryos *in situ* (not straightened). If more than one embryo is present, measure several and give an approximation of their size (e.g., 5 embryos, CR = 16 mm). Some collectors make it standard practice to denote the number of embryos in the right (R) and left (L) horns of the uterus (e.g., 5 embryos; 3L, 2R). In mammals having more than one embryo, some may die and be resorbed into the uterus. Resorbed embryos appear conspicuously smaller and underdeveloped when compared with normal ones. Be careful to distinguish any resorbed embryos when recording embryo counts (e.g., 5 normal embryos: 3L, 2R, CR = 15 mm; 2 resorbed embryos: 1L, 1R, CR = 3 mm). Embryos can be easily preserved by placing them in a vial of 10 per cent neutralized formalin. If embryological studies are contemplated, preserve in Bouin's solution (see section 6).

The uterus should also be examined for the presence of placental scars. In some mammals (shrews, rodents, carnivores), after a female gives birth placental scars form at sites in the uterine wall where embryos were implanted. These scars appear as yellow to black pigmented spots on the inside of the uterus (Fig. 17B). Although the scars become increasingly paler and smaller with age, they may persist to one year in mice and rats. Generally the number of scars corresponds to the number of embryos. However, embryos that died during pregnancy will also leave scars on the uterine wall and because scars from several litters may be present, the number of scars is not always an accurate indication of litter size. Nevertheless, the presence or absence of placental scars is important to determine the reproductive history of the animal. If scars are readily visible, count them and record the number in each horn of the uterus. Where scars are badly faded, scars from several litters are present, or when scars are obscured by embryos, it may not be possible to count them accurately. If two (or more) sets of scars representing two (or more) different pregnancies are present, one set of scars will appear larger than the others. Although it may be impossible to count all scars, it is important to indicate that two (or more) sets of scars were observed. Examples of placental scar data: 2 sets of placental scars present—not counted; 4 placental scars (3L, 1R); no placental scars present.

With the data obtained from examining the uterus, females can be classified as follows: nulliparous—no embryos or placental scars; primiparous—embryos *or* one set of placental scars; multiparous—embryos and one (or more) sets of placental scars present, *or* two (or more) sets of placental scars present.

4.6 Locality Descriptions

Mammalogists are increasingly concerned with studies of geographic variation and

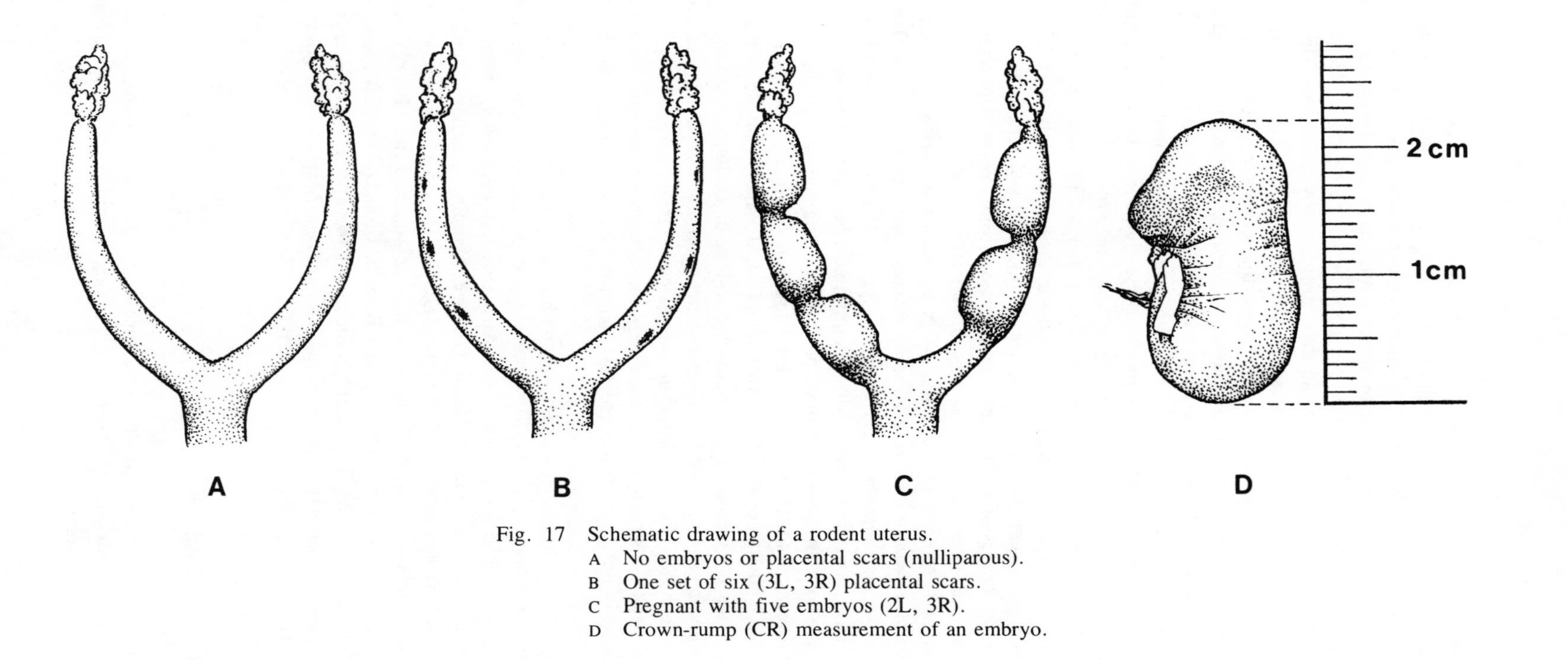

Fig. 17 Schematic drawing of a rodent uterus.
A No embryos or placental scars (nulliparous).
B One set of six (3L, 3R) placental scars.
C Pregnant with five embryos (2L, 3R).
D Crown-rump (CR) measurement of an embryo.

variation within populations. For this research detailed and accurate descriptions of collecting sites are required. The following locality data are required: country, state or province, county or equivalent, township or equivalent, local area or equivalent (town, lake name, name of nearby mountain), and the *precise* latitude and longitude to the nearest minute. Unfortunately, accurate mapping of collecting sites is frequently prohibited by insufficient data provided by collectors. Examples of problems resulting from incomplete and inaccurate locality descriptions are: locality description consisting only of a village name that cannot be located on any map, locality description consisting of the name of a town or village (no other data) when there may be several towns of the same name in the country, failing to distinguish between road distance (speedometer mileage) and map or airline distance.

To avoid such problems, please follow these guidelines for documenting your collecting localities. If possible give the state or province and the county (or equivalent if such exists) in the country. This is helpful in limiting the search for a locality. The local area designation (a mountain, lake, river, valley) should be given when it is a prominent feature. To locate a collecting site precisely, give map or airline distance (miles or kilometres) and direction from a permanent reference point and latitude and longitude to the nearest minute if possible. A town or a mountain peak is a suitable reference point if it appears on small-scale maps (e.g., 1:500 000). If the name of a town or village is the same as a nearby mountain, river or lake, insert the word town in parentheses, for example, Rainy River (town). Remember that a good reference point must be readily identifiable now as well as any time in the future. Transitory features that may be moved, renamed, or eliminated such as wayside taverns, small ponds, roads, highway numbers or junctions, and campgrounds should be avoided. Also avoid local names that do not appear on published maps for they may be impossible to locate. Distance from the reference points should be given in map or airline distances and not in road distance calculated from a speedometer. Road mileage in mountainous terrain is extremely inaccurate. Also, new road construction and reroutings make highway distances transitory. If road mileage is used, you must indicate that the distance is road mileage and not airline distance.

Finally, we urge collectors to include with their field catalogues and field notes large-scale topographic maps (1:50 000 or 1:250 000) with collecting sites indicated. Maps of various scales for Europe, Asia, Australia, Latin America, Oceania, and the United States may be obtained from Defence Mapping Agency, Department of Defence, Topographic Center, Washington, DC 20315, USA. Topographic maps of Canada and an index listing available maps may be obtained from Energy Mines and Resources Canada, Surveys and Mapping Branch, Canada Map Office, 615 Booth Street, Ottawa, K1A 0E9. Topographic sheets for the Province of Ontario may be obtained from Ontario Ministry of Natural Resources, Map Office, Queen's Park, Toronto, M7A 1W4.

4.7 Habitat Descriptions

A description of the habitat in which the specimen was collected gives an indication of its ecological distribution. Therefore, provide as much habitat information as feasible for all specimens.

A useful habitat description includes: (1) an indication of the general biome (desert, arctic tundra, savannah, coniferous forest, hardwood forest, alpine tundra, rain forest). (2) The dominant or most prevalent types of plants (palm grove, white spruce and balsam-fir forest, banana plantation, acacia scrub forest). If you know the scientific names of plants, use them in the description. (3) Note the elevation, for example, 3000 m above sea level (ASL). Elevation can be accurately estimated from topographic maps or with a pocket altimeter. For calculations from maps, use the units given on the map rather than converting metres to feet or vice versa. (4) Any pertinent information on the history of the area should be given (area burned over by forest fire in 1968; area cut over for logging in 1970). (5) Any pertinent information on soil type or geological formations.

Rare mammals that are seldom captured by collectors warrant particularly detailed descriptions of their habitat. Frequently little, if anything, is known about the ecology of these species. For bats, describe roosting sites or netting sites (roosting in humid cave, roosting in palm tree, netted over small stream).

Examples of informative habitat descriptions are as follows: trapped in alpine zone, 400 m ASL, vegetation scrubby willows and mosses; netted in dense mature forest, 1000 m ASL; from a large colony roosting in humid limestone cave, near banana plantation, 400 m ASL; trapped in treeless, grassland area 200 m ASL; trapped in mature jackpine balsam-fir forest. Figure 7 illustrates typical habitat data recorded in field notes.

4.8 Methods of Capture

Describe how the specimen was collected (trapped, shot, netted, found dead, road kill, poisoned). If all your specimens were captured in the same manner, then write this at the beginning of your catalogue or field notes to avoid repetition.

4.9 Miscellaneous Field Notes

This includes any miscellaneous behavioural or ecological observations that you may have made. Although these observations may seem to be trivial, they frequently prove to be valuable in natural history studies. Examples are as follows: netted at 08 00 h; trapped between 08 00 h and 09 00 h; unusual colour phase; observed copulating with field number 20; specimen is young of field number 17; species observed feeding on papaya fruit; males and females roosting in different parts of the cave; tick on right ear; this colony of bats observed flying out of cave at 18 00 h; many of this species occupying the same burrow. Also record observations on climatic conditions in your field notes. Weather conditions may influence small mammal trapping or the netting of bats. Examples of these observations are: rainy season, dry season, heavy rains during trapping period.

4.10 Photographic Records

Colour slides or black and white photographs of specimens and collecting sites are a useful way to supplement documentation data. Photographs of rare mammals that are seldom captured by collectors or aberrant specimens (e.g., unusual colour phases) may prove invaluable, particularly if these specimens represent species new to science. Close-up photographs of the facial regions of live mammals taken in mist nets or live traps are an excellent way to illustrate the structure and colour of fleshy appendages (e.g., bat nose-leaves) that may fade or shrink in study skins or fluid-preserved material. You may also wish to photograph mammals that are prepared as complete skeletons without skins in order to make a permanent record of pelage. Photographs of carcasses of beached whales and dolphins prepared as skeletal material are also valuable. You will enhance your habitat descriptions (section 4.7) with photographs of the vegetation on trap lines or at bat netting sites. Photographs may be the only way to document rare or endangered species protected by law.

Most field biologists prefer the 35 mm single-lens reflex camera because of its light weight and versatility. Various accessories (bellows, extension tubes, macrolenses) are made for close-up work. The authors have found a wide-angle lens (35 mm or 28 mm focal length) ideal for habitat photographs.

5. *Methods for Preparing Specimens*

Three kinds of museum specimens are usually prepared from mammals: skins with accompanying skulls and/or partial skeletons, complete skeletons, and entire mammals preserved in fluid. Each of these has advantages and disadvantages and the kind of specimen prepared depends on the objectives of the collector. Ideally, a representative series of a species from a given locality should contain all three types of specimens. Study skins are essential for analysing pelage colour and moult patterns; fluid-preserved specimens are valuable for studying anatomy and histology. Skeletons are useful for studies in comparative anatomy, geographic variation, and determining age but skeletons of many species are poorly represented in collections. Collectors are encouraged to prepare complete skeletons (at least one male and one female) for each species collected when feasible.

The condition of the specimen will often determine the kind of specimen that should be prepared. For preserving in fluid, live mammals from mist nets or live traps are most suitable. Keep mammals alive in collecting bags or cages until they are to be prepared. Mammals from snap traps or specimens that have been frozen can usually be prepared as study skins. Mammals from snap traps should be prepared as quickly as possible because they decompose rapidly. Once the skin of a mammal begins to "slip" (fur falling out), it is difficult to salvage a satisfactory study skin from it. Decomposed specimens in which the internal organs have deteriorated and the fur is slipping are best prepared as skeletons. A "skull only" should be salvaged if the remaining carcass is badly damaged.

5.1 Preserving in Fluid

Because changes in tissues occur shortly after death, specimens should be preserved immediately after killing. To effectively kill small mammals (bats and mice) without damaging the skin and skull, use an airtight jar, can, or plastic bag with a wad of cotton containing a few drops of chloroform or ether. Avoid inhaling these chemicals as they are toxic to humans. Ether and chloroform are also highly inflammable. Larger mammals (hares, foxes) can be humanely and quickly killed by injecting them in the heart region with Euthanyl (sodium pentobarbital). The recommended dosage is 1 ml per 2.3 kg (5 lb.) body weight. Used by veterinarians, Euthanyl is a restricted drug that can only be obtained by prescription. You may find it necessary to calm large mammals with ether or chloroform before injecting them.

The preparation of entire mammals in fluid involves two steps: fixing the tissues of the specimen with a solution such as 10 per cent formalin, Bouin's solution, or sodium acetate and transferring the specimen for permanent storage to a preserving fluid, for example, 65 to 70 per cent ethanol or 45 to 60 per cent isopropyl alcohol. ''Fixing'' halts enzyme processes in tissues and hardens or ''sets'' the specimen. Preservatives prevent the growth of microrganisms and also prevent gradual chemical or physical changes in the specimen's structure.

Unless specimens are to be stored for two months or longer before shipment, the collector need be concerned only with fixation. Fluid-preserved specimens received from field collectors are washed for 12 to 24 h in water and then transferred to 65 per cent ethanol for permanent storage in museum collections. The best fixative for the field collector is a solution of 10 per cent buffered neutral formalin. Formalin is a solution of formaldehyde. The commercially available formalin is usually a 37 per cent (weight/weight) or 40 per cent (weight/volume) solution of formaldehyde. These commercial formaldehyde solutions are treated as 100 per cent formalin; therefore, to produce a 10 per cent formalin solution, mix one part 40 per cent formaldehyde with nine parts water. To reduce volume, most collectors carry full strength (40%) formaldehyde solutions in the field and dilute them to 10 per cent formalin before preserving their specimens. A wide-mouthed glass jar or a plastic pail makes a convenient container in which to dilute formaldehyde and to fix specimens. Because formalin solutions are usually acidic (pH 3.0 to 4.6), they tend to decalcify teeth and excessively harden tissue. To neutralize acidity, add a teaspoon of powdered borax (sodium tetraborate) or a tablespoon of household ammonia to 1 gal. (3.8 L) of 10 per cent formalin. Best results are obtained by using a precise mixture of salts to buffer the formalin to neutrality (pH 7.0). For example, a mixture of 4 g of acid or monobasic sodium phosphate monohydrate ($NaH_2PO_4H_2O$) and 6.5 g of dibasic sodium phosphate anhydrate (Na_2HPO_4) will buffer 1 L of 10 per cent formalin. About 40 g of this salt mixture neutralizes 1 gal. (3.8 L). This dry, salt mixture can be prepared in advance and carried into the field in plastic bags. Other equipment for preserving specimens includes a hypodermic syringe and hypodermic needles for injecting the body cavity and larger muscles. If injection equipment is not available, mammals can be preserved by making a slit in the abdomen to open the body cavity to allow the formalin to enter.

After specimens have been killed, weighed, measured, and assigned a number (see section 4), a field tag should be tied securely to each. If you use paper labels, be certain that they will not disintegrate in the preserving fluid. Using waterproof ink or a pencil, write the number and sex symbol (♀ for female and ♂ for male) on both sides of the label.

Once properly tagged, lay the mammal on its back and, with the syringe full of formalin, insert the needle into the abdomen and slowly fill the body cavity until it becomes turgid. Do not inject too much fluid, but be sure that the body cavity is full and firm. For large mammals insert the needle into the larger muscles and inject a small amount of the formalin. For mammals with fleshy tails (rats) slit the tail with several cuts using a scalpel or sharp knife.

After injection, the specimen is fixed by placing it in a jar, pan, or pail containing 10 per cent buffered neutral formalin. Care should be used to keep the specimen in a normal, relaxed position for it will retain this shape permanently once it has been fixed. If the mouth is not locked tight, prop it open with a small piece of wood or a piece of cotton before the specimen is fixed. This permits examination of the teeth for identification without damaging the mouth parts. For bats, wings should be partially closed in a natural position. Usually the wings are in satisfactory shape if the bat is killed in a relaxed position. Do not overcrowd specimens in a container before they are thoroughly fixed (usually 12–48 h). Two factors are important: (1) to keep the specimen in a natural undistorted position; (2) to make certain there is sufficient volume of formalin to properly fix the specimen. A safe rule is to have sufficient formalin to completely cover the specimens.

After 12 to 48 h specimens are fixed and they may be packed more tightly in containers for storage. Avoid using metal containers (unless acid-proof lined) as formalin causes almost immediate rust and corrosion that will discolour specimens. If only metal lids are available, a waxed, cardboard liner will seal the container for only a short time because of corrosion. Therefore, place a piece of waxed paper or a sheet of plastic over the mouth of the jar before screwing on the lid. If specimens are to be stored for a long period before shipment (more than 8 weeks), they should first be washed thoroughly in fresh water and then placed in 65 to 70 per cent ethanol or 45 to 60 per cent isopropyl alcohol.

5.2 Preparing Skins

Three types of skins can be prepared for museum specimens: (1) Traditional study skins that are filled with cotton or a similar material to approximate the natural shape of the mammal. As the leg bones are ordinarily left in the skin, a complete skeleton cannot be obtained and usually only the skull is kept. (2) Flat skins, which consist of the skin stretched over a cardboard outline. For a flat skin, it is possible to obtain a skull and most of the skeleton. (3) Tanned skins, which are first dried and later tanned for permanent preservation. A skull and skeleton can be obtained from a mammal prepared as a tanned skin with only the terminal digits left on the skin.

The choice of skin depends on the size of mammal and the collector's objectives. Small mammals (bats, rodents, insectivores) are prepared as study or flat skins. The study skin has been traditionally used by museum collectors. This type of skin may be more time-consuming to prepare, especially for the inexperienced, but it is invaluable for studying pelage. Flat skins have the advantage of being quick and easy to prepare and they provide both a skull and a skeleton. However, flat skins are difficult to compare with study skins when analysing pelage. Skins from mammals larger than a fox are too bulky to be made into study or flat skins on cardboard and must be prepared as tanned skins. Study skins can be prepared for small fur-bearing mammals up to the size of a fox, however, some collectors prefer to make tanned skins for these mammals.

A. Study Skins

Begin with a midventral incision from the level of the last rib to near the anus (Fig. 18). Always cut to one side of the penis or vagina so that the external genitalia remain attached to the skin. To keep the skin clean and dry, cornmeal, borax, magnesium carbonate powder, or sawdust may be sprinkled on the skin or placed in a skinning tray or pan to absorb blood and body fluids. Another skinning technique is to make an incision that extends across the lower abdomen and down the inside of the leg to the heels (Fig. 18). Leave the external genitalia on the skin by cutting between the anus and genitalia. For males that have a baculum in the penis, be careful not to cut or damage this structure for it may be an important aid in identification. With mammals such as bats or mice, the baculum can be left intact in the penis to dry on the skin. If bacula studies are anticipated, remove the entire penis and store it in 100 per cent glycerine, 10 per cent formalin, or 70 per cent alcohol. For larger mammals, the baculum should be extracted from the penis, tagged and dried with the skull and any skeletal material.

With fingertips, a scalpel handle, or blunt forceps, work the skin free of the body wall in the vicinity of the incision. Try not to cut into the body cavity. Holding the hind foot, push the knee joint upward towards the midline of the body. Peel the skin off the leg to the ankle, then sever the hind leg at the hip or knee joint with scissors or a scalpel (Fig. 19). When the hind legs are free, work the skin to the base of the tail. Use care while skinning around the anus and the anal scent glands found in some mustelids (Mustelidae).

If the mammal has a nonfleshy tail (e.g., some bats), cut it close to the trunk and leave the tail vertebrae in the skin. If a fleshy tail is present, slip it out of the skin and later replace it with a wrapped wire. Rolling the tail on a table top or skinning board will help loosen the connective tissue that attaches the tail vertebrae to the tail sheath. For shrews, mice, and other small specimens grasp the tail at the base of the sheath with the thumb and index finger (Fig. 20) of one hand. Press the thumb- and finger-nails firmly against the tail vertebrae. Then with the other hand, slowly pull the tail vertebrae until they are free of the skin. If the tail vertebrae break off in the tail sheath, you must split the skin of the tail and remove the vertebrae. After inserting a tail wire, the incision should be sewn with a fine needle and thread. For larger mammals, it may be necessary to hold the base of the tail sheath with heavy forceps or two blocks of wood. For mammals larger than a squirrel, it is usually necessary to split the tail by a longitudinal incision in order to remove the vertebrae.

With the tail free, the skin can now be peeled back to the region of the front legs (Fig. 21). Do not pull the skin off the body, as this will result in an overstretched study skin. A recommended method is to use one hand to gently push the skin off the body and with the scalpel held in the other hand, sever any connective tissue holding the skin to the carcass. Remove the skin from the front legs down to the ankle. For bats, detach the skin as far as the elbow joint. With scissors or a scalpel, cut the front legs (or wing bones of bats) at the shoulder joint (Fig. 22). Peel the skin over the chest area to the base of the skull.

Probably the most difficult stage of the skinning operation is to remove the skin from the head region without damaging the ears, eyelids, lips, and skull. Use a sharp scalpel for skinning the head region. Carefully work the skin over the head until you reach the cartilaginous bases of the ears (Fig. 23). Pick away any fatty tissue that may

Fig. 18 Initial incision is made midventrally (dashed line) or across the heels (solid line).

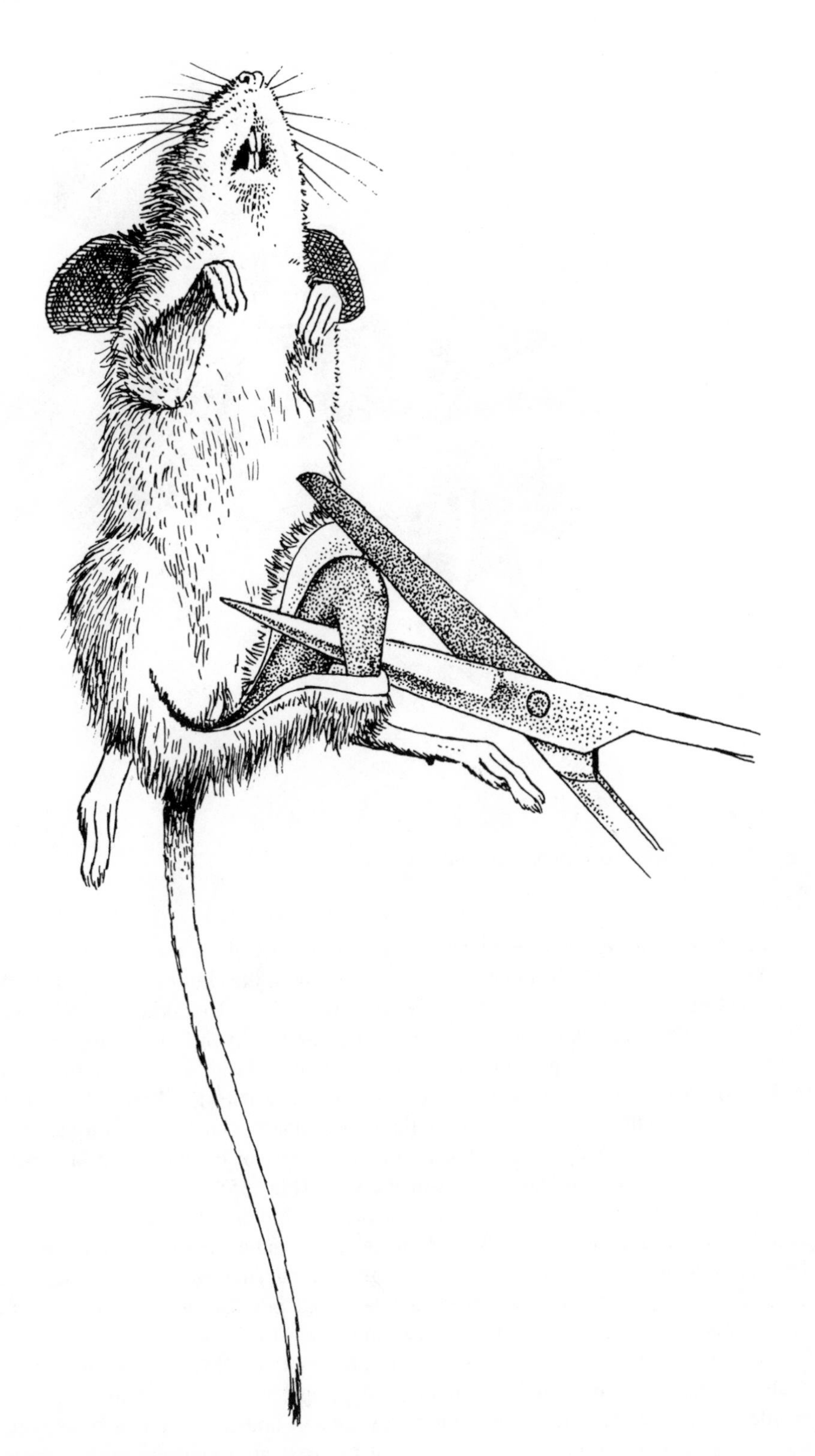

Fig. 19 Severing the hind leg at the knee joint.

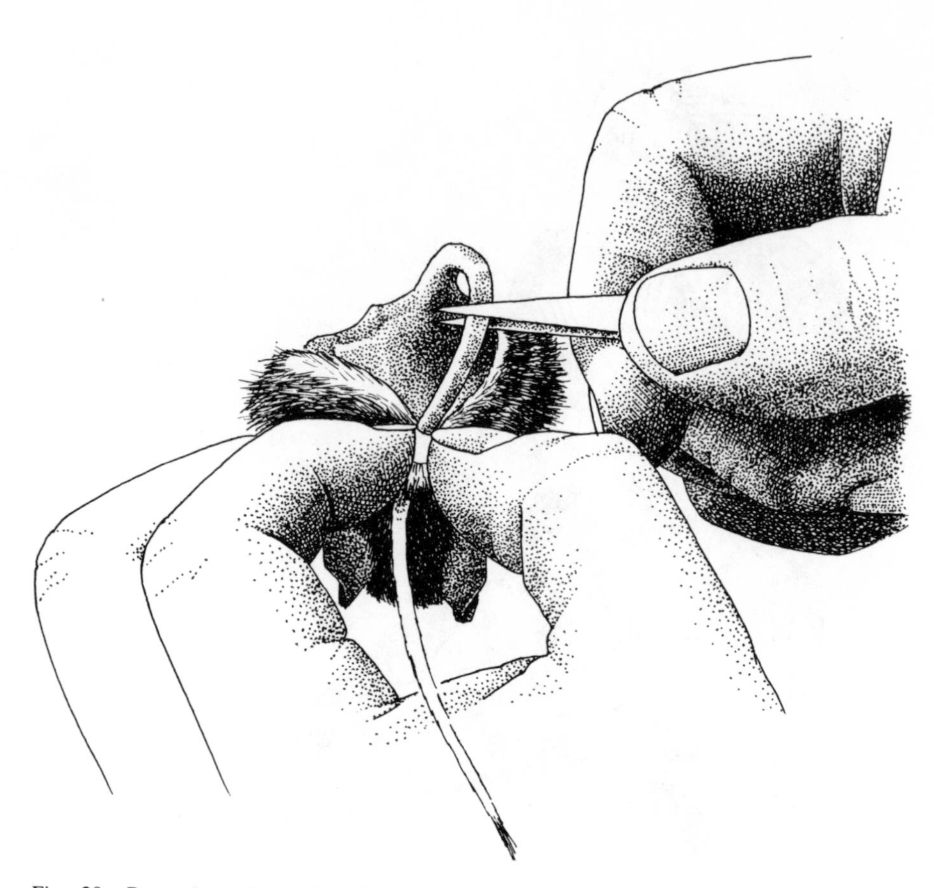

Fig. 20 Removing tail vertebrae from the tail sheath.

obscure the ear cartilage and sever the cartilage at the base of the ear. Continue to peel the skin over the head until the eyes are exposed. With the skin held away from the head, cut the membrane that covers the eyes (Fig. 24). The skin should be still attached in the eye region at the front corner of the eyelid. Carefully cut this attachment with your scalpel but avoid cutting into the eyelid, as the skin of the eye region will tear when the skin is stuffed. Work the skin to the lips and sever the connective tissue that attaches the lips to the skull. Finally, peel the skin forward until it is attached to the body only at the tip of the nose. Cut the nasal cartilage being careful not to cut into the nasal bones of the skull (Fig. 25).

Once the skin is removed, dissect the carcass for reproductive data (section 4.5) then direct your attention to the skin. Remove all fat and excess flesh from the skin. To accelerate drying and inhibit insect damage, rub a drying agent into the flesh side of the skin. At the ROM, we use magnesium carbonate powder for study skins and flat skins. This powder is available from most biological or chemical supply companies. Borax can also be used as a drying agent-preservative. However, there is some evidence that borax may affect red-coloured pelage; therefore, try to keep borax powder off the fur. Arsenic, arsenic and borax, arsenic and alum, alum and potassium nitrate, and arsenic soap have been used in the past as drying-preserving agents. Because alum may affect fur colour and arsenic is toxic, these chemicals are not

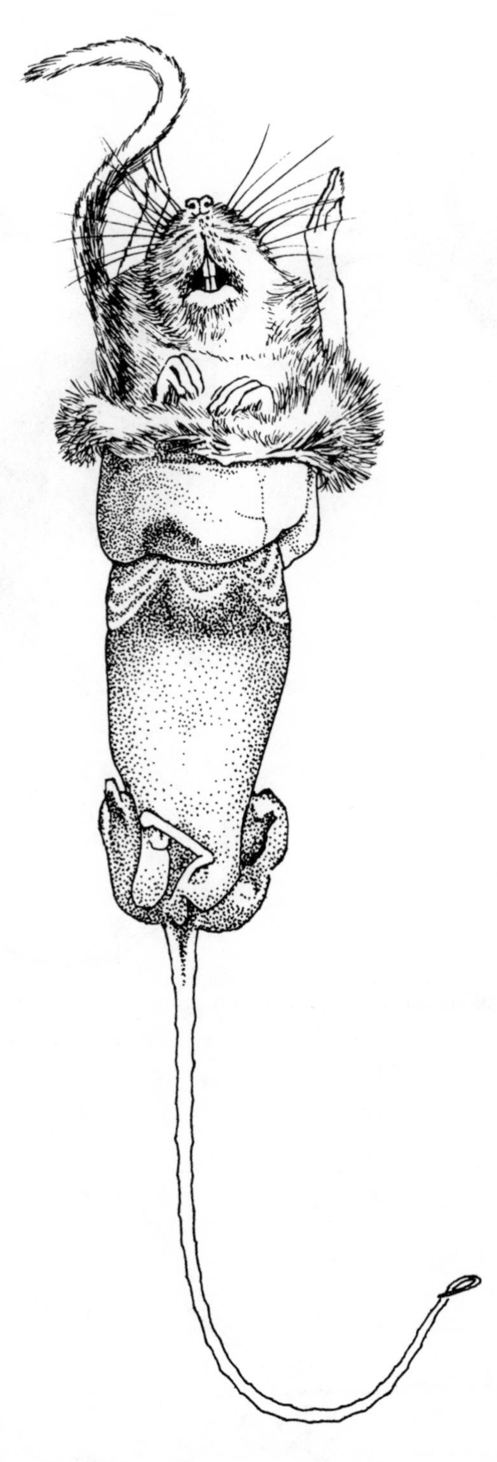

Fig. 21 Skin being peeled off the carcass.

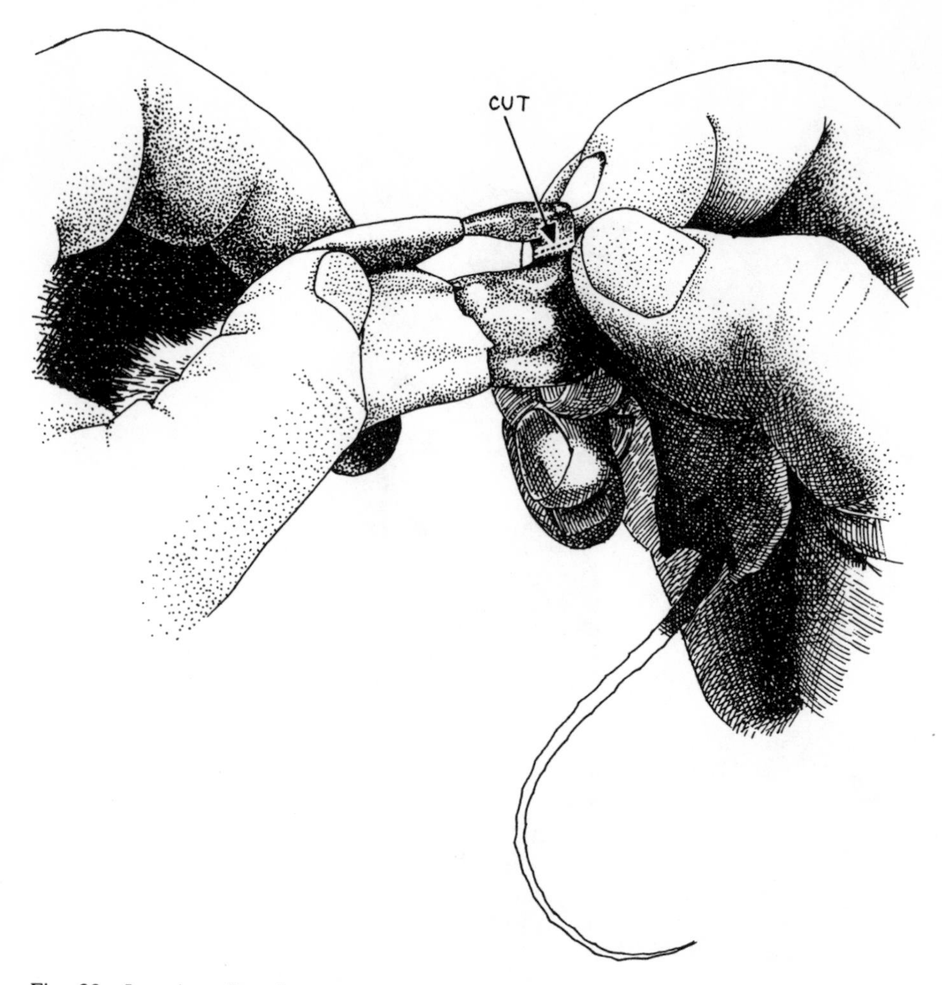

Fig. 22 Location of cut for severing the front legs.

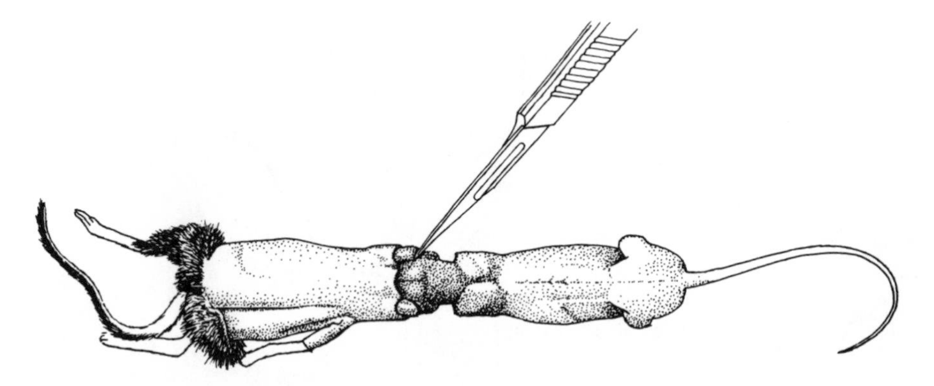

Fig. 23 Cutting the cartilage at the base of ears.

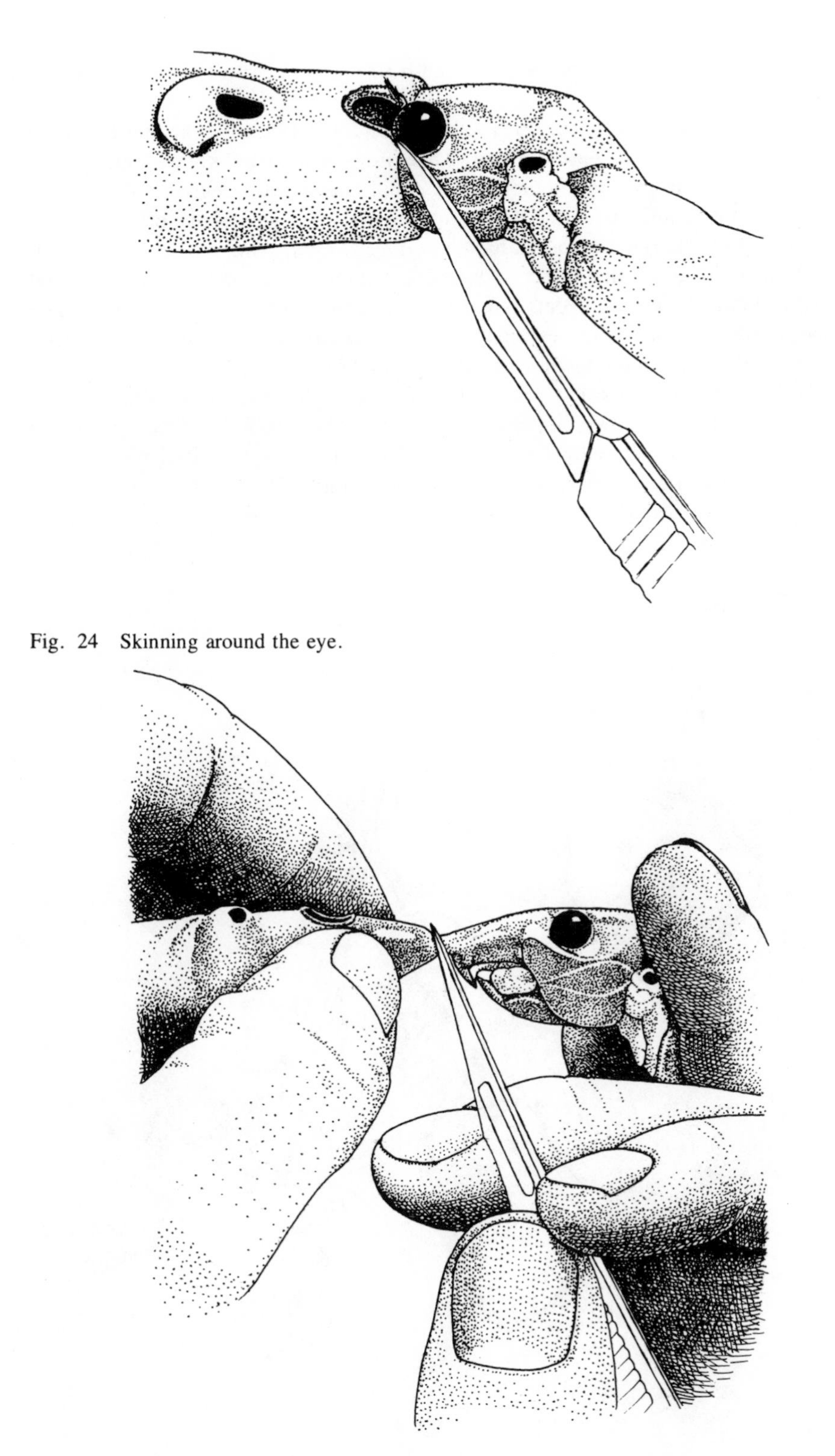

Fig. 24 Skinning around the eye.

Fig. 25 Removing the skin from the nose.

recommended and they should only be used if magnesium carbonate or borax is not available.

For smaller mammals (mice, shrews, and bats), fat and flesh can be picked off the skin with your fingers; however, it may be necessary to use a dull knife to remove this material from skins of larger mammals. A simple method for degreasing skins that have heavy fat deposits is to dip them in naphtha or white gas. Shake off any excess gas and roll the skin in sawdust to hasten drying. Do not wring the skin as this will stretch it. A liberal dusting of fur with magnesium carbonate or sawdust before and during skinning usually prevents blood from adhering to the skin. Heavily matted blood around wounds may be removed with water or alcohol on a cotton wad, followed by dusting with magnesium carbonate or sawdust.

Remove the muscle tissue from the leg or wing bones with scissors, scalpel, or forceps (Fig. 26) and rub the bones in magnesium carbonate. Restore the legs to their approximate original shape by wrapping the bones with cotton to replace the muscles. With the skin still reversed, sew the lips together (Fig. 27).

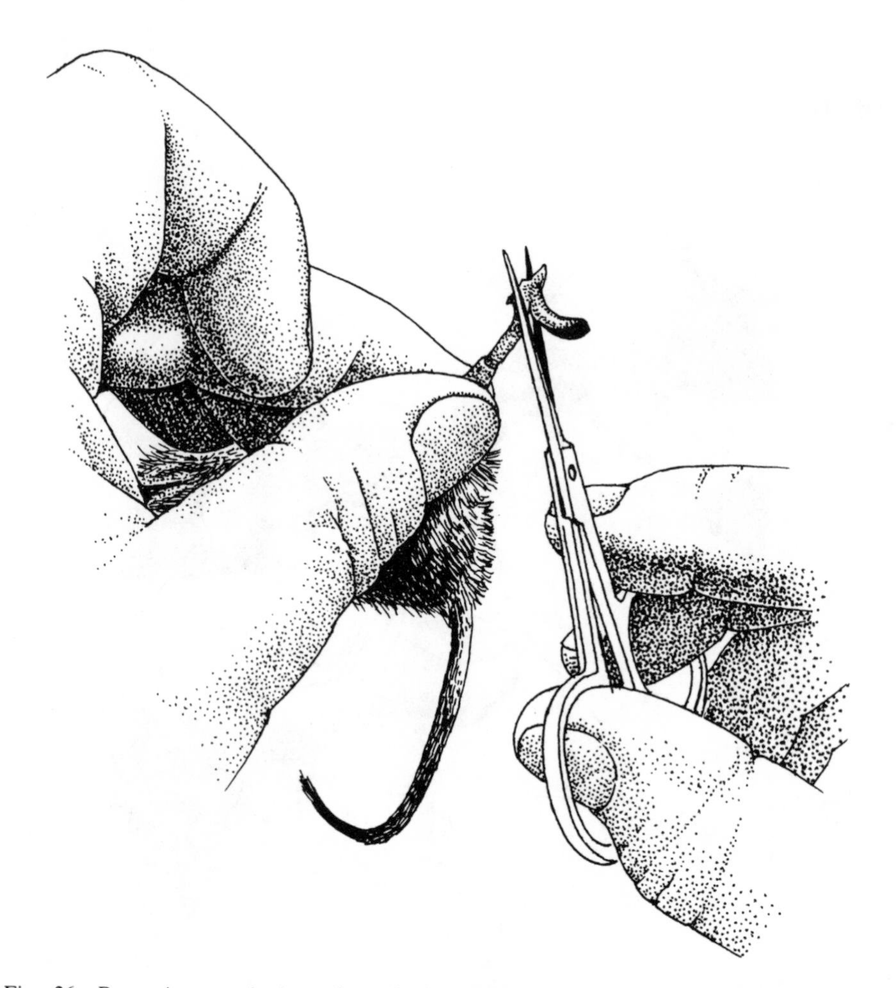

Fig. 26 Removing muscle tissue from the leg bone.

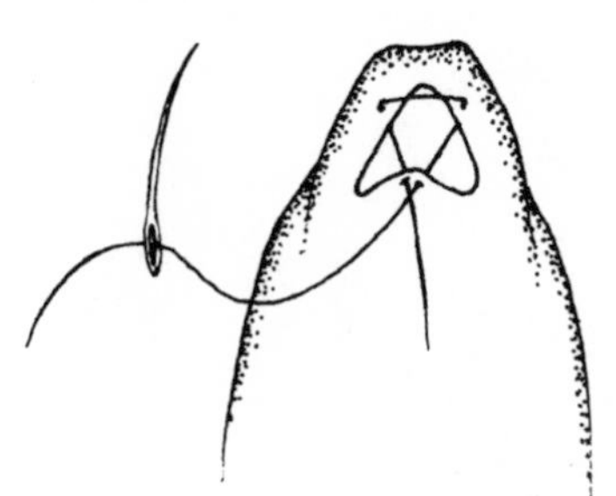

Fig. 27 Stitch used to close the lips.

Now the skin is ready to fill with a body and head made from a single piece of cotton. Some collectors prefer to use fine tow rather than cotton. Roll the cotton into a smooth cylindrical bundle that is slightly longer and thicker than the body of the mammal (Fig. 28). Form the head region of the cotton filler with a pair of forceps by pressing in the centre at the end of the roll (Fig. 29). Grasp the two corners on either side of the forceps, fold together, and take a new hold of the pointed end and shape as a smooth cone (Fig. 30).

Place the cone into the head of the skin and reverse the skin over the points of the forceps (Figs. 31, 32). Adjust the eyes, ears, and mouth, then continue to reverse the skin slowly over the cotton until the specimen is completely filled. The length of the cotton body can be trimmed with scissors to fill the skin properly (Fig. 33). For study skins of small species of bats, mice, and shrews, some collectors prefer to construct a separate head rather than use a single piece of filling. Best results are obtained by constructing the head from fine tow and wrapping it with a thin wisp of cotton. Use the skull as a guide for size. Instead of sewing, hold the lips in position by inserting small pins into the tow. After the head is shaped, fill the body with a single piece of cotton.

If the tail vertebrae were removed, then an appropriately sized wire wrapped with cotton must be inserted into the tail for support (Fig. 34). The collector will require the following gauges of wire for preparing skins: 12 gauge (hares), 16 gauge (large squirrels), 18 and 20 gauge (small squirrels), 22 and 24 gauge (mice, rats, shrews), 26 gauge (small shrews and small bats). If available, use Monel wire as it does not corrode. Cut the wire to a length that extends from the tip of the tail to midway into the body. A loop at the body end of the wire provides added strength and stability to the finished specimen. Wrap thin wisps of cotton to form a shape similar to that of the original tail vertebrae (Fig. 35). It may be necessary to moisten the tail wire with saliva to make the cotton adhere. Species with long tails that taper to a fine tip (jumping mice), present a problem because the tail tip is too narrow to accommodate the usual tail wire. A very thin wire (26 gauge) can be tapered to a fine point with a file. However, a better method is to prepare tapered wires in advance of fieldwork by dipping precut wires into an acid solution to form a long, tapering point that will fit the tail sheath. Dipping Monel wire for 5 min in a solution of one part hydrochloric acid (HCL) and two parts nitric acid (HNO_3) will effectively produce a tapering point. Oxide deposits on the wire can be removed by placing wires for several minutes in an enamel tray containing hydrochloric acid. Pad the portion of the looped

Fig. 28 Rolling cotton into a cylindrical bundle.

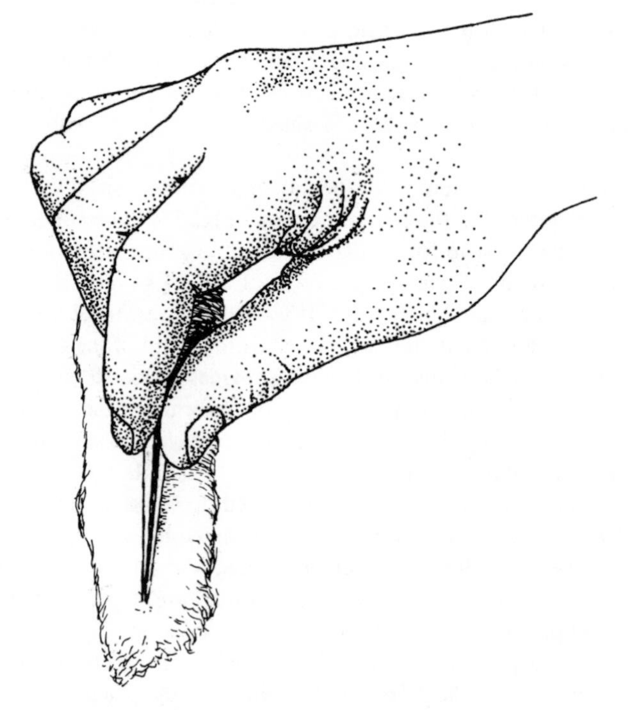

Fig. 29 Grasping the cotton with forceps.

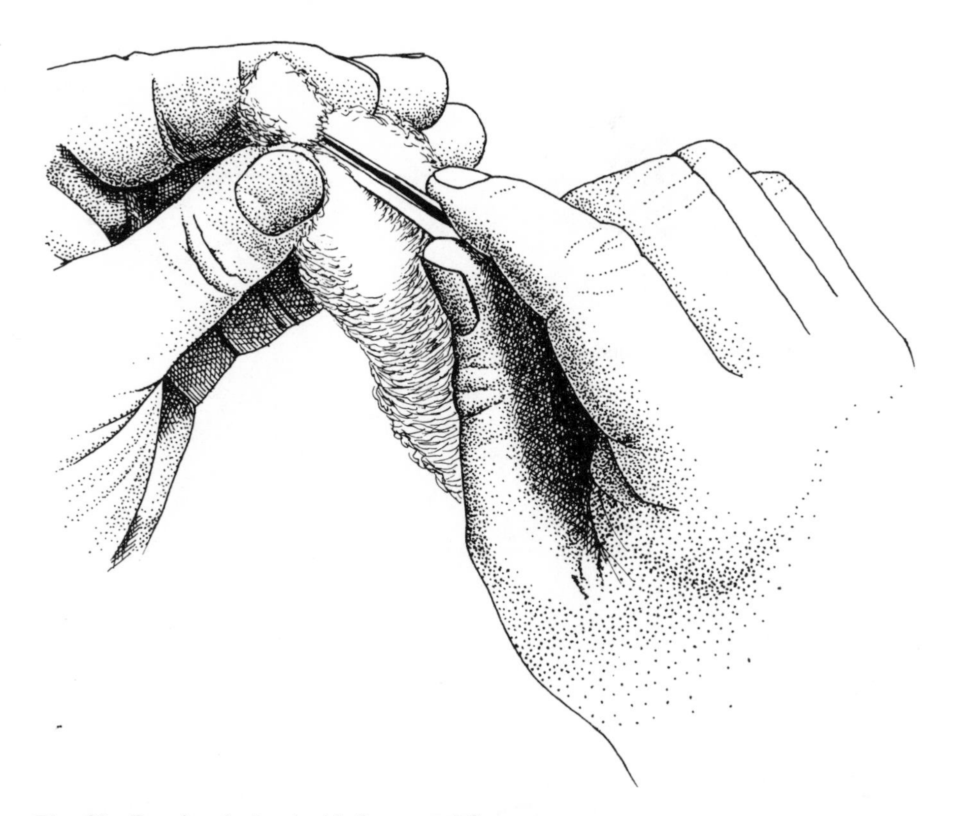

Fig. 30 Forming the head with fingers and forceps.

tail wire that extends beyond the tail into the body cavity with a thin piece of cotton then stitch the midventral incision with a fine needle and thread (Fig. 36) and tie a field tag (see section 4.2) to the hind foot of the skin.

The next step is to anchor the study skin to a pinning board (cardboard, corkboard, Styrofoam, or wallboard) for drying. Careful pinning is the key to a well-prepared skin. For most mammals the front and hind feet are positioned parallel to the body and held in place with pins through each foot and a pair of pins at the outer side of each hind foot near the heel (Fig. 37). Anchor the tail by a pair angled across its base and by one pair angled across the tip. To shape the ears and head, use pins placed against the side of the skin.

Check to be sure the head is symmetrical and, if necessary, a thin (insect) pin may be used to anchor it in place. The eyelids may be held open by pulling through a small bit of cotton from the head. A final check of the specimen should be followed by cleaning the fur with a small brush (a toothbrush works well) to remove dirt or dust (Fig. 38).

A recommended method of pinning bat wings (Fig. 39A) is to place a sharp pin (insect pins are preferable, but any sharp pin will do) in the wing joint near the thumb. Position the wing so that the forearm is nearly parallel to the body and the upper arm joins the body in a natural position at a slight angle. Some of the membrane above the forearm and upper arm should be exposed, but the wings should not be overstretched. The second pair of pins is inserted at the elbow to hold the forearm in the desired

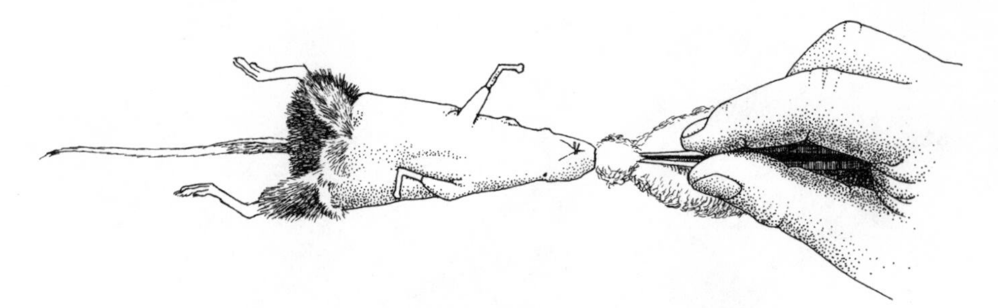

Fig. 31 Cotton body is inserted into the head region.

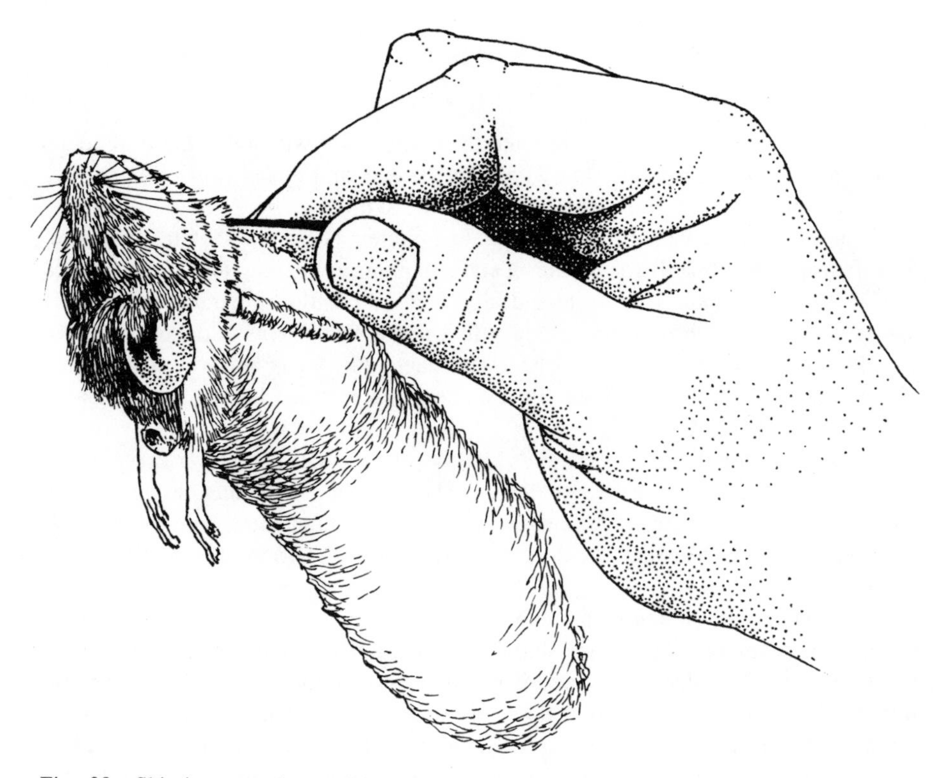

Fig. 32 Skin is reversed over the cotton.

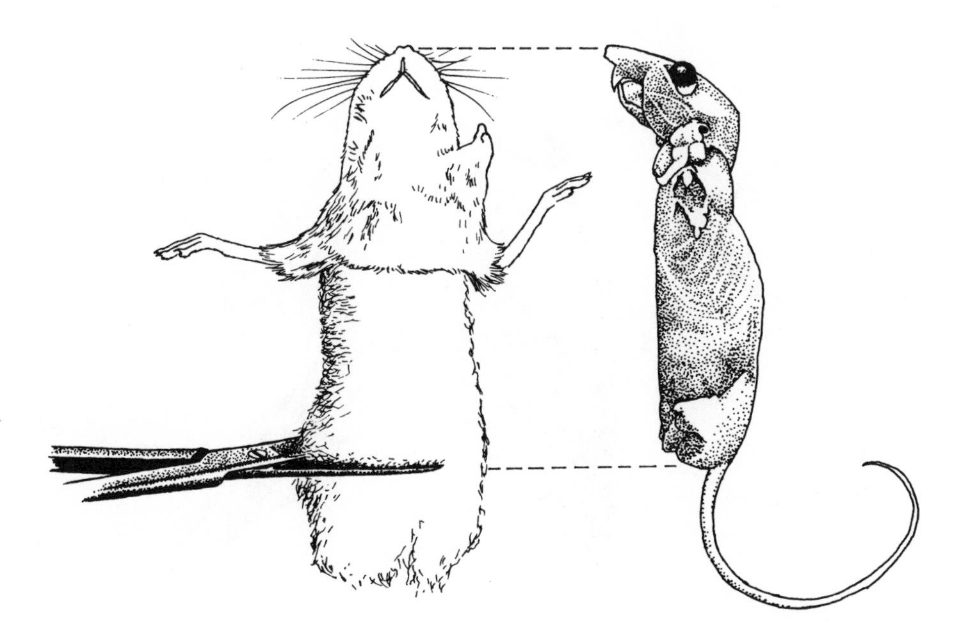

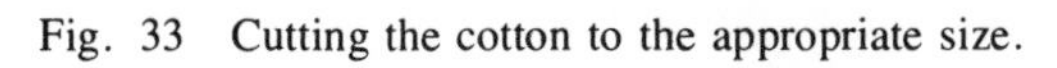

Fig. 33 Cutting the cotton to the appropriate size.

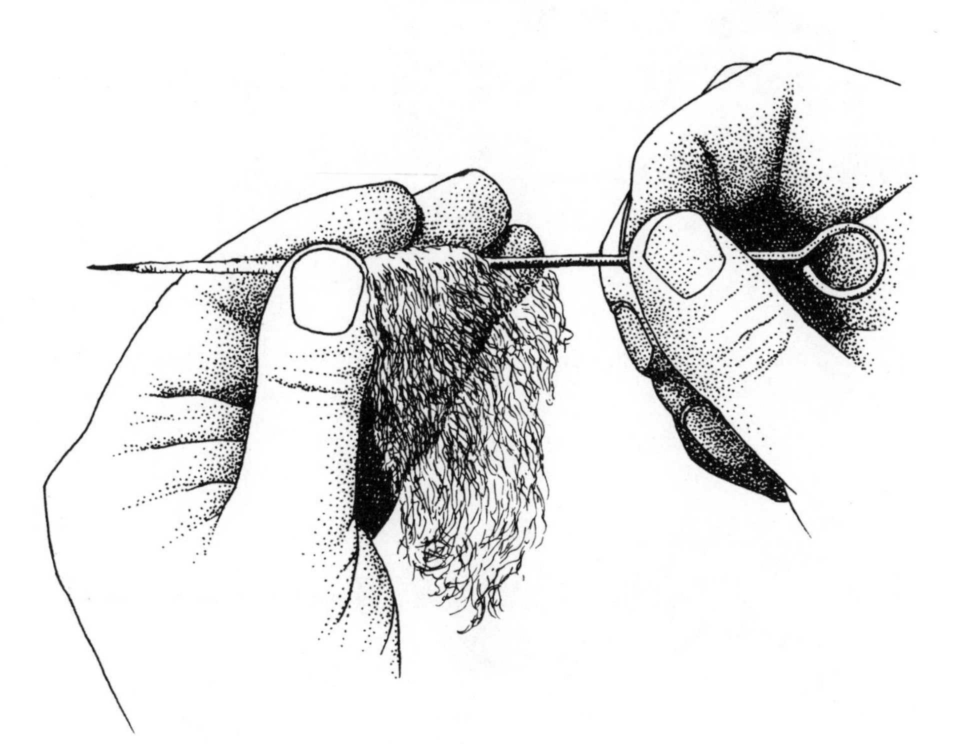

Fig. 34 Wrapping the tail wire with cotton.

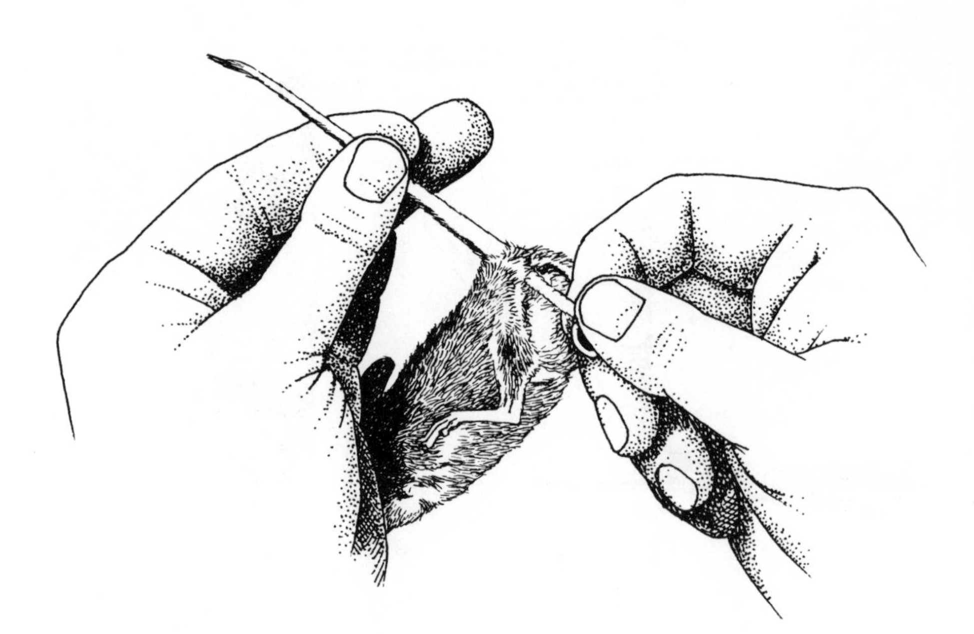

Fig. 35 Inserting the wrapped wire into the tail sheath.

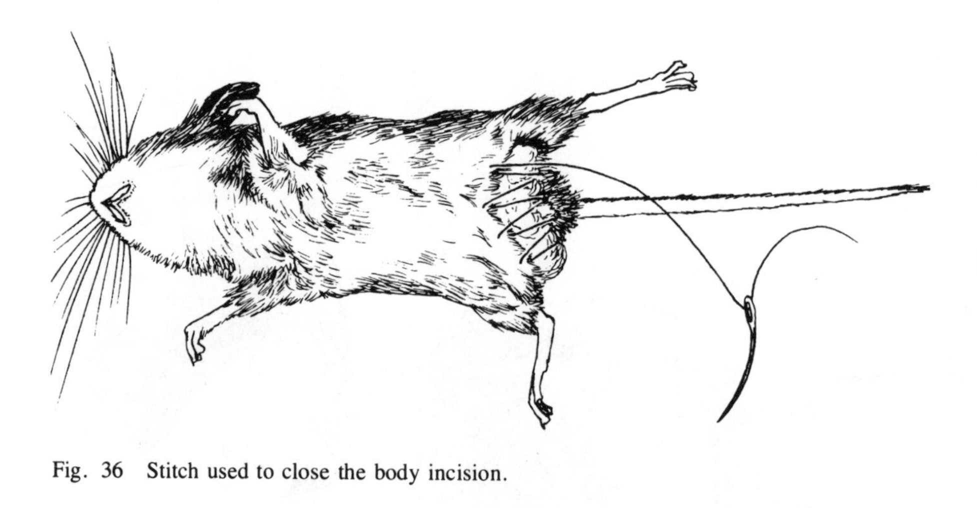

Fig. 36 Stitch used to close the body incision.

Put feet under head- pointing fwd

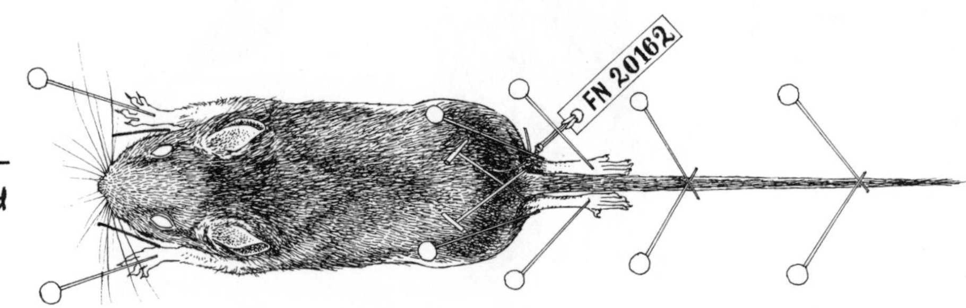

Fig. 37 Skin is pinned out for drying and a field tag is attached to hind foot.

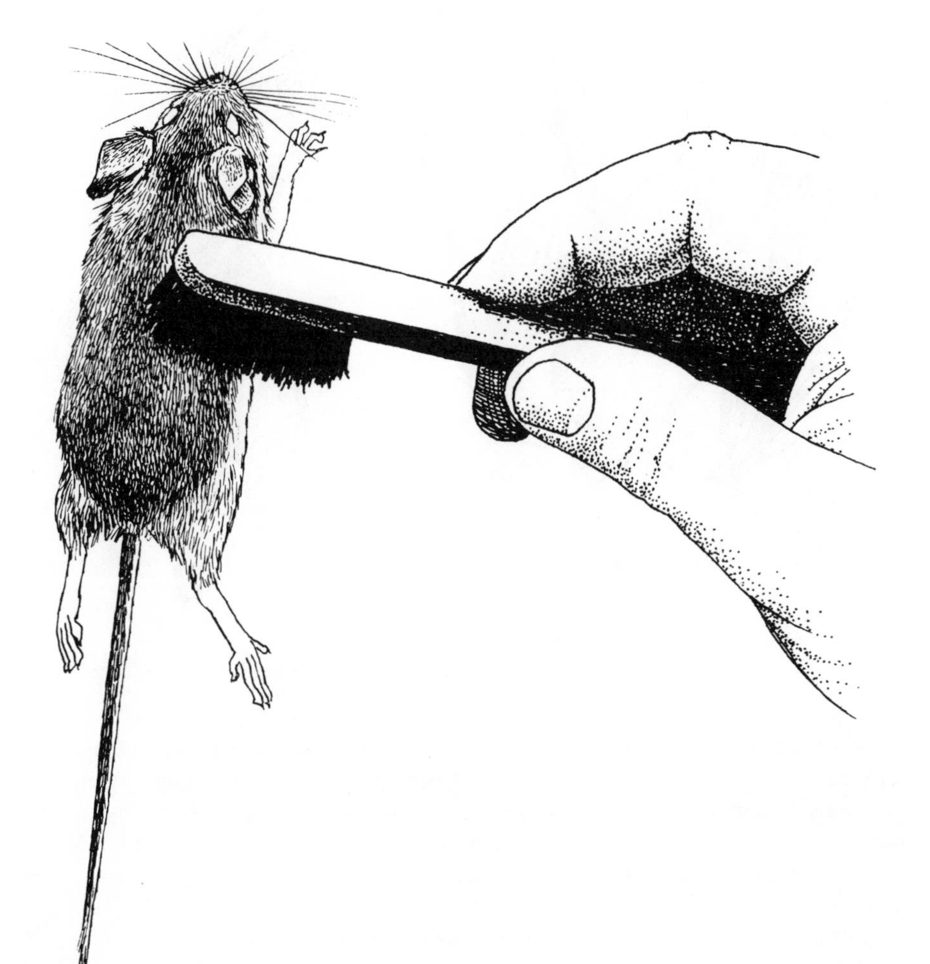

Fig. 38 Brushing the fur with a small brush.

position. A third pair of pins may be required to hold the end of the upper arm in the appropriate position close to the body. Pin each foot, pulling the body into the desired position. If a tail is present, extend and anchor it with a pin at the tip and with crossed pins over the base of the tail near the body. If a membrane is present between the tail and legs, extend and anchor it by placing pins near the end of the calcar (the cartilaginous structure extending along the edge of the membrane from the heel of the foot). The wing bones are then pinned so that each digit or finger is spread slightly away from adjoining ones.

The pinned specimen must be dried thoroughly before shipping or transporting. Drying time will vary considerably with local conditions. In hot dry climates, stuffed skins may dry in one day; however, in humid climates it may be extremely difficult to dry skins completely. A shaded area with good air circulation provides the best

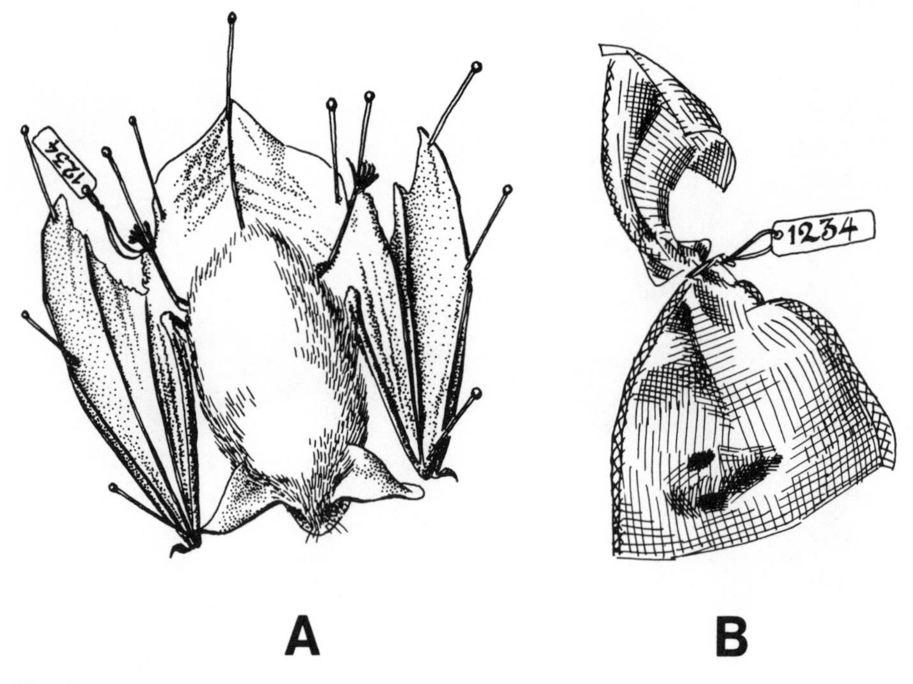

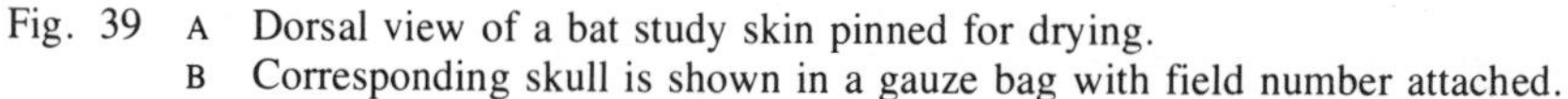
Fig. 39 A Dorsal view of a bat study skin pinned for drying.
B Corresponding skull is shown in a gauze bag with field number attached.

conditions. Do not place skins in direct sunlight as this fades pelage and intense heat may cause excessive shrinking of skins. Under poor drying conditions, it may be necessary to unpin specimens and to expose their undersurface to the air by turning them upside down. Take care not to bend or break the ears when doing so. When skins are dried, remove the pins and store them in chests or shipping boxes (see section 7).

Ants and egg-laying flies can cause extensive damage to skins and most collectors protect their drying skins by storing them in special wooden cases that have screened openings for ventilation. A properly screened drying chest should be part of your standard field equipment; however, a temporary drying container can be constructed in the field using a few pieces of wood and cheesecloth. At the ROM we use wooden drying chests (90 cm × 50 cm × 35 cm; 35 inches × 20 inches × 14 inches), that contain 80 cm × 42 cm (32 inches × 17 inches) sheets of 2.5 cm (1 inch) Styrofoam. The sheets of Styrofoam are separated by 2.5 cm (1 inch) high, wooden frame spacers. For small mammals (shrews, mice), a single spacer is sufficient for separating Styrofoam sheets, for larger mammals (hares), two or three spacers are required. Study skins are pinned to the Styrofoam for drying. When they are sufficiently dried, pins are removed and each layer of skins is covered with cotton.

B. Flat Skins

Flat skins mounted on cardboard can be prepared for such species as shrews, mice, squirrels, and small carnivores. Many collectors use the method described by

Anderson (1965) where both front and hind feet are left on the skin. We recommend the following procedure because it enables the collector the obtain both a flat skin and a skeleton from the specimen. The skeleton is complete except for the one front and one hind foot that are left on the skin.

Rather than making a cut along the midline of the abdomen, begin the cut at one heel, cutting through the skin at the back and inner side of the leg, across the base of the tail between the anus and external genitalia, and then extend the incision to the opposite heel (Fig. 18). Leave the external genitalia on the skin. Detach the skin from the legs and cut through the leg bone at the ankle of only one leg; leave that foot on the skin. On the opposite leg, detach the skin to the ankle and then cut the skin there. This foot is left attached to the body of the mammal as part of the skeleton. Now remove the skin from the tail and pull it towards the front legs. Do the same for the front legs as the hind legs—leave one foot on the body of the mammal and detach the other foot with the skin. Remove the skin from the head region being careful not to damage the ears or lips. Dissect the carcass for reproductive data (section 4.5). Clean all fat and excess flesh from the skin and, to hasten drying, rub a drying agent (borax, magnesium carbonate) into the flesh side of the skin (see section 5.2A).

Unlike the conventional study skin, the flat skin is stretched on a piece of cardboard or corrugated pasteboard. To prepare the stretcher, lay the skin flat on the board and trace its outline. Shape the card with scissors, leaving a sufficient amount of the card behind the shaped outline to support the tail and to permit writing of field data (Fig. 40). Use a board sufficiently thick to support the skin fully. Cardboard sheets of different thicknesses should be carried into the field by collectors who intend to prepare flat skins. Pull the skin over the stretcher board, being careful not to overstretch. For small mammals, the skin is put on the board, fur side out, but for mammals larger than a squirrel, the skin should be stretched on the board flesh side out for a few days. Then, after the skin has partly dried, reverse it and pull it on to the board, fur side out.

Insert a wrapped tail wire for support as with conventional study skins (see section 5.2A) and tie the hind foot and base of the tail to the board with thread (Fig. 41). You will need heavy sail-makers' needles for piercing heavy cardboard. Small pins are used for holding the front foot in position for drying and for shaping the lips if necessary. Use a toothbrush for a final cleaning of the fur. Tie the field number (see section 4.2) to the stretcher card and write the field number on the card in case the tag is lost (Fig. 41).

A modified flat skin is used by museum collectors for hares and rabbits (Fig. 42). Because the hind legs are supported on a wooden stick, this type of skin is sturdier than the conventional study skin. Also it requires less space for storage. A heavy stretcher (corrugated cardboard) is cut to fit the skin. A wooden stick which extends from the head to beyond the hind feet is attached to the board by means of cord or wire. Wrap the cardboard stretcher with a thin layer (0.5 cm; 0.25 inches) of cotton and insert it into the skin. It may be necessary to trim the posterior end of the board with shears to fit the skin properly. Wires should be inserted in each leg to provide support. Secure the hind legs to the stick with wire or heavy thread; the front legs are positioned parallel to the head. The ears can be held together flat against the skin in a natural position with a single stitch of thread. Sew the skinning incision and tie a field number to one hind leg. At the ROM, we are using a modified version of this method for our hare skins. The skin is prepared in the same manner as the traditional hare flat

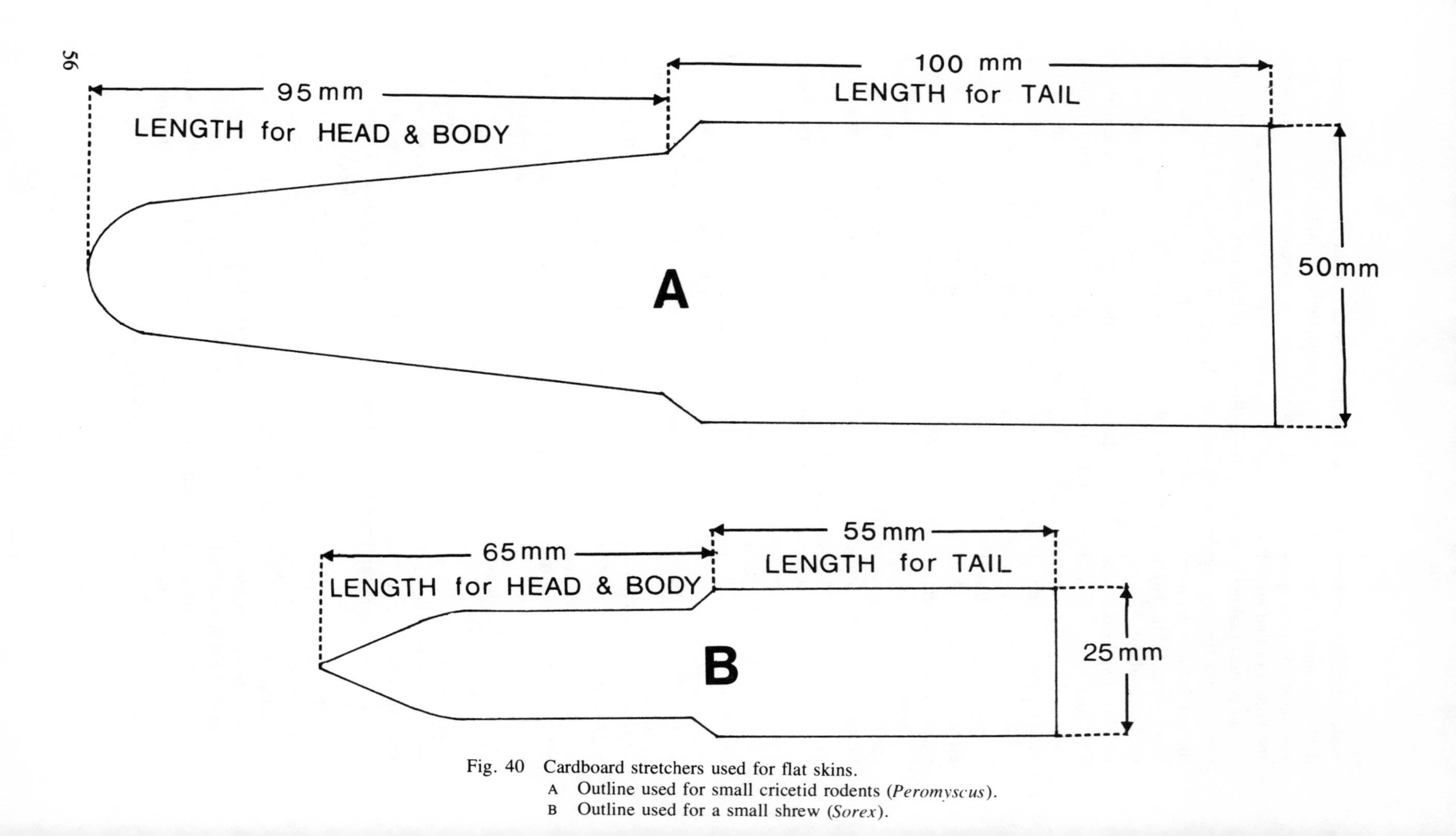

Fig. 40 Cardboard stretchers used for flat skins.
A Outline used for small cricetid rodents (*Peromyscus*).
B Outline used for a small shrew (*Sorex*).

skin except that only one hind foot and one front foot are left on the skin. The other feet are left on the carcass as part of the skeleton.

Flat skins dry quickly and under optimum conditions they may be sufficiently dried in 24 h. Follow the precautions for drying study skins, that is, keep flat skins out of direct sunlight and protect them from insect pests. After they have been dried, flat skins can be packed compactly into boxes for shipping.

C. Skins to be Tanned

Mammal skins larger than a fox must be prepared for tanning, although smaller fur-bearing mammals (foxes, ermine, beaver) may also be prepared for tanning. Skinning pelts for tanning may be either "cased" or "open". Pelts from fur-bearers other than beaver are usually prepared "cased" by professional trappers; however, the choice of skinning method is really a question of personal preference. Large mammals such as bears, deer, moose, and seals should be skinned "open". For a discussion of techniques for mammals that require special treatment, see Anderson (1965).

"Cased" skins are removed from the mammal in much the same manner as the flat skin (see section 5.2B). Make an incision from foot pad to foot pad along the hind legs. Detach the skin from the hind legs to the foot. The skin can be separated from the carcass by pushing down with thumb and fingers between the skin and carcass. Skin out the feet leaving only the claws on the skin and remove the tail vertebrae from the tail sheath. For mammals larger than a squirrel, split open the tail on the ventral side. With the tail vertebrae free, pull the skin down the carcass to the front legs. Detach the skin from the front legs, leaving the claws on the skin, then pull the skin over the head region. Carefully skin around the ears, eyes, mouth, and nose.

Cased skins should be stretched, fur side in, over a frame for drying. For the collector who prepares the occasional pelt, a stretcher can be improvised from soft wood, wire, or corrugated cardboard. However, if you plan to collect a number of

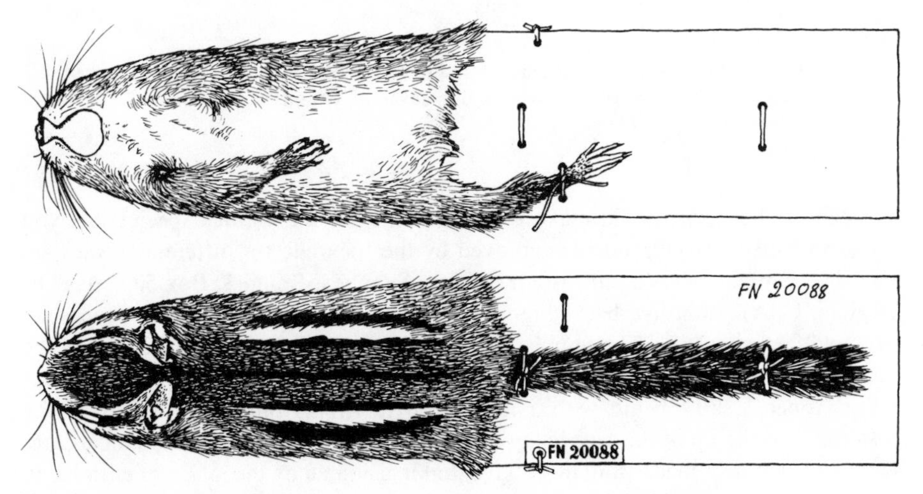

Fig. 41 Example of a flat skin prepared from a chipmunk (Sciuridae).

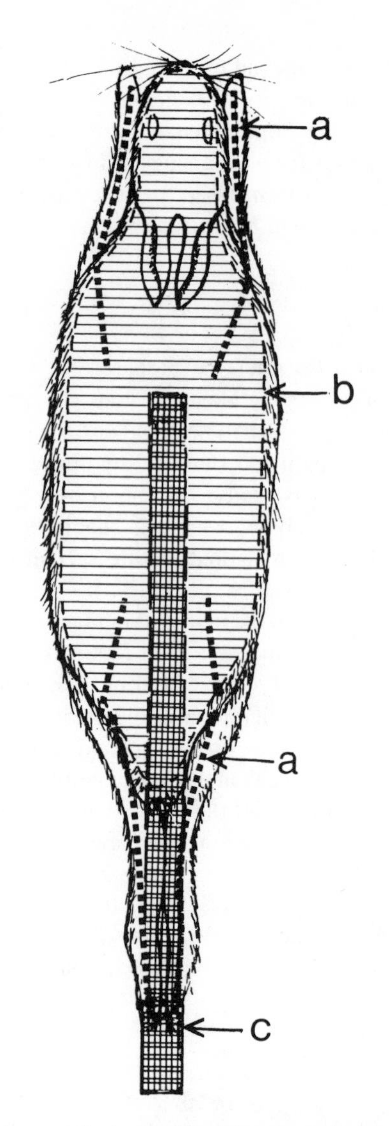

Fig. 42 Modified flat skin used for hares. Hind feet are tied to a stick which is attached to a cardboard outline covered with a thin layer of cotton: a leg wires, b cardboard outline, c wooden stick.

fur-bearers, then you should make a drying board for the various species that you intend to collect. Drying boards approved by the fur trade for different species are available commercially (Southeastern Outdoor Supplies, Route 3, Box 503, Bassett, Virginia, USA). Adhesive-backed templates in various standard sizes and shapes are also sold commercially. The template is placed on a board which is then cut to the appropriate size and shape.

For "open" skins begin with a midventral incision from the throat extending posteriorly to the tip of the tail, being careful to cut to one side of the genitalia (Fig. 43A). Cuts are then made from this midventral incision along the inside of each leg to the foot. Usually the claws or hoofs are left on the skin; however, if the carcass is to

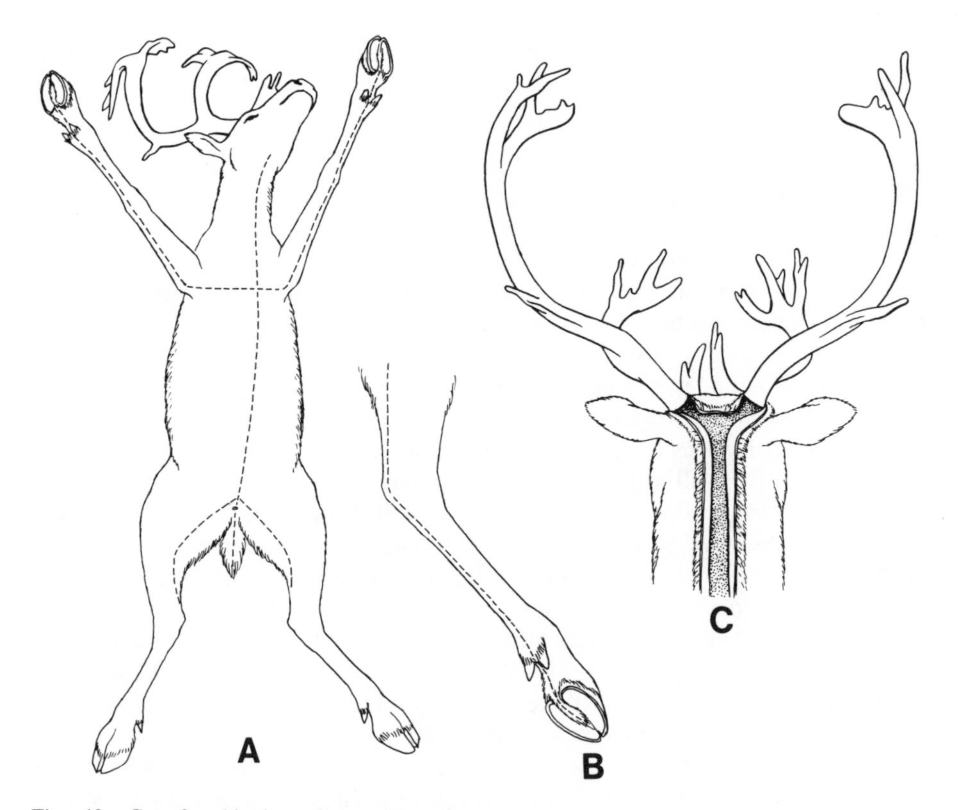

Fig. 43 Cuts for skinning a large mammal.
A Ventral view of cuts.
B Enlarged view of cuts for skinning leg and hoof.
C Cuts for skinning around antlers.

be prepared as a skeleton, then leave only one hind foot and one front foot on the skin—the other feet remain on the carcass. The skin of some mammals (deer and caribou) can be pulled off easily by hand, but for other mammals (bears), the skin must be removed by careful cutting. Figure 43C illustrates the cuts that should be made for skinning around horns or antlers.

Once the skin is off, remove as much flesh and fat as possible and sponge off any blood from the fur. Be extremely careful when defleshing. If too much of the skin is removed, the fur will fall out when the pelt is tanned. An ordinary dull table knife makes a good fleshing tool. Spread a layer of salt (sodium chloride) evenly on the entire flesh side of the skin and rub it in thoroughly with the hands. Stretch the skin out, flesh side exposed, and allow it to dry for about 24 h (longer in humid conditions). Cheesecloth or mosquito netting can be used to keep egg-laying flies off the skin. After 24 h, shake off any water and excess salt. Resalt the skin, then fold the head and legs in and roll the skin into a bundle. The skin should be periodically checked for signs of decomposition and the presence of fly eggs or larvae and it may be necessary to apply another treatment of salt. The skin must be dried as thoroughly as possible prior to shipment or transport. After the skin is received from the collector, museums usually send it to a commercial tanner for tanning. It is important *not* to treat the skin with chemical preservatives other than salt in the field as these may interfere with the tanning process.

5.3 Preparing Skulls

As soon as the study skin is prepared, direct your attention to the skull. Care should be taken to prevent damage to any part of the skull. Separate the skull from the carcass by severing at the joint of the skull and the first vertebra (atlas). Fragile skulls from small mammals (shrews, mice, small bats) are dried without any cleaning. However, the brain, eyes, tongue, and heavy muscle layers should be removed from skulls about the size of squirrels or larger. A piece of wire with a small hook on the end can be used to pick the brain tissue out of the cranium, or the brain tissue can be flushed out of the skull with a syringe. Cut the muscles attaching the eyes and tongue with scissors and then pull these organs from the skull using forceps. Heavy muscle tissue can be removed by using a scalpel or scissors, but be extremely careful not to damage the thin processes on the skull. For skulls with antlers in velvet (deer, moose, etc.), it may be necessary to split the velvet with a knife to facilitate drying.

Thoroughly dry skulls in the field. An efficient technique for drying skulls is to place them in cloth bags. Gauze skull bags are usually supplied to ROM collectors; however, if you do not have these bags, similar ones can be constructed from cheesecloth or other porous material. Write the field number in pencil on a heavy paper or cardboard tag and tie it to the skull. The same number can be attached to the outside of the bag (Fig. 39B). Skull bags can be strung on a wire and put out in a ventilated place to dry. Ensure that skulls cannot be reached by animals that might be attracted to them. When using a vehicle in the field, an excellent technique for drying skulls is to tie wire strings of skulls securely under the hood where the heat of the engine and air flow during travelling quickly dries them. It may be necessary to store skulls inside the vehicle at night to protect them from such predators as racoons or cats. After skulls have completely dried, they can be compactly packed for shipping. As with study skins, the time required to dry skulls varies with climatic conditions. Before packing, ascertain that there is a skull for every skin and note in your catalogue any damaged skulls.

5.4 Preparing Skeletons

If the specimen is to be prepared as a flat skin or tanned skin, then remove the skin by the methods described in section 5.2. However, if the mammal is decomposed with the fur slipping, remove and discard the skin in the quickest manner possible, being careful not to damage the skeleton. Because wing membranes must be removed from the bones of the hand, bats prepared as skeletons should be skinned as follows. Detach the skin from the legs as far as the ankle, then cut the skin with scissors at the ankle joint. Peel the skin from the carcass until you reach the wings. Detach the skin from the arm of the bat; then by slowly pulling towards the wing tips, peel the wing membrane off the wing bones, being careful not to break the delicate bones. Remove the skin from the head region in the usual manner (see section 5.2).

When the skin is off the carcass, dissect for reproductive data (see section 4.5), then extract all internal organs from the body cavity. Any organs left in the carcass will decompose rapidly. Next, remove the larger muscle tissues from the bones. The amount of defleshing required depends on the size of the mammal. To avoid damaging the delicate skeletons of such small mammals as shrews, mice, or small bats, do not attempt to cut off the small amount of flesh. Simply leave the skull attached to the skeleton and allow the entire carcass to dry. For medium-sized mammals (squirrels, hares, fruit bats, small carnivores), remove the muscles with scissors and forceps. Although it is not necessary to disarticulate the skeleton for these mammals, that is, to cut the ligaments that hold the bones together, you must separate the skull from the skeleton and extract the brain, tongue, and eyes (see section 5.3). Considerable work is involved in defleshing the skeletons of large mammals (deer, bears, seals) and a sharp skinning knife is essential for cutting the heavy muscles and ligaments. You must at least partially disarticulate large mammals in order to reduce the skeleton to manageable sections. Separate the skull and remove the brain, tongue, and eyes (see section 5.3).

After defleshing, skeletons must be dried thoroughly before shipping or transporting. The best method for drying skeletons of small- to medium-sized mammals is to place them in gauze bags that protect them from egg-laying flies but provide ventilation for rapid drying. If you do not have these bags, make them from cheesecloth or similar material. Tie a field number to the skeleton and place a single skeleton in each bag. For bats tie the wing bones against the body to prevent them from breaking off. After tying the top of the bag tightly with string, hang it to dry following the precautions described for drying skulls (see section 5.3). Burlap bags make excellent drying bags for skeletons of large mammals. Each bag should have the field number securely attached. Two important rules to follow when preparing skeletons are: (1) do not place skeletons in plastic bags, as they will decompose instead of drying; (2) do not treat skeletons with any chemicals as they will inhibit the activity of dermestid beetles that are used in many museums to clean skeletons.

6. *Special Techniques*

6.1 Karyotyping

Slides of somatic chromosomes can be prepared in the field using the *in vivo* bone marrow technique. The method is simple, quick, and produces slides of good quality for conventional staining. Most workers use chromosomes from tissue culture for banding studies.

A large metal toolbox is invaluable for carrying karyotyping equipment in the field (see Appendix 2). Sodium citrate solutions will support growth by bacteria and consequently must be prepared fresh each day. Weighing and separating the required daily quantities of sodium citrate crystals before a field trip saves considerable time. Small vials or plastic bags are suitable containers for sodium citrate. Although slides can be stained in the field, staining solutions add considerable bulk and weight to the karyotyping kit. Slides can be stained with good results 1 to 2 months after preparation in the field. Most hand centrifuges hold four centrifuge tubes; therefore one can process four specimens at a time. Although the volume of suspension in centrifuge tubes could produce a dozen or more slides, we generally prepare four to six slides for each specimen.

Practise the procedure in the laboratory before attempting to do it in the field. This will enable you to become familiar with the various steps and to verify the quality of results.

Mammals that have been karyotyped must be kept as voucher specimens (study skins or preserved in fluid), for chromosome slides without voucher specimens are virtually worthless.

The following technique modified from Baker (1970) has been used by ROM staff to karyotype bats, small rodents, and small carnivores in the field. A list of equipment required for karyotyping is given in Appendix 2.

1. Dilute one or more vials (10 mg) of Velbe (Eli Lilly & Company, Indianapolis, Indiana, USA) with distilled water to produce a 0.025 per cent solution. One method is to take a serum bottle containing 100 ml of bacteriostatic sodium chloride (Abbott Laboratories, Montreal, Canada) and remove and discard 20 ml of sodium chloride solution with a sterile syringe. Dissolve the contents of two 10 mg vials in the serum bottle to produce an 80 ml stock solution of Velbe. Keep this refrigerated. For a working solution, transfer 10 ml of stock solution to any empty Velbe container. If working with colchicine, use a 0.04 per cent solution.

2. With a 1 ml tuberculin syringe, inject the mammal intraperitoneally. The dosage of Velbe is 1 unit (0.01 ml) per gram of body weight. Use the same dosage with colchicine. After injection, the mammals must be kept alive for 1.5 to 2 h. Bats can be kept in individual collecting bags; rodents or carnivores in cages.

3. After 1.5 to 2 h, anaesthetize the mammal, dissect out the humerus (bats) or the femur (rodents, carnivores) being careful not to damage the proximal end. Remove muscle tissue from the bone.

4. Using a 5 ml Luer Lock syringe, flush out bone marrow with 3 ml of 1 per cent sodium citrate solution (made fresh daily). The size of the syringe needle will depend on the size of the bone. Generally, no. 21, 24, or 26 needles are required for small mammals.

5. Vigorously break up and suspend cells with a pipette. Then let the solution stand for 5 to 20 min (more than 20 min may rupture cells).

6. Centrifuge at 500 to 1500 rev/min for 2 to 4 min.

7. With a pipette remove and discard the supernatant fluid being careful not to disturb the clump of cells.

8. Slowly add one pipette of Carnoy's fixative (three parts absolute methanol:one part glacial acetic acid) to the centrifuge tube. By carefully adding the fixative down the inside of the tube with a pipette, you do not disturb the clump of cells. With the tip of the pipette close to but not touching the cells, carefully extract the fixative. Add fresh fixative, suspend the cells and let stand for 10 to 12 min.

9. Centrifuge for 2 to 4 min, discard supernatant, then resuspend the cells in fixative.

10. Repeat this procedure two or three times.

11. After the final centrifugation, resuspend the cells in fixative.

12. With a pipette, place two or three drops of suspension on a clean microscope slide. Ignite by quickly passing a match over the slide. Do not touch the slide with the match. Some workers prefer to air dry slides. Use microscope slides with frosted ends and record the field number of the specimen on the slide with a lead or diamond pencil.

13. Slides can now be stained in the field or sealed in a slide box and later stained in the laboratory. To prepare the stain, mix eight parts of warm distilled water with one part Giemsa stain (filter the Giemsa before mixing with water). Slides are stained for 13 min in a Coplan jar. About 50 ml of stain are required to fill a Coplan jar. The destaining process requires five steps in Coplan jars: (1) rinse in acetone; (2) 1 min in acetone; (3) 1 min in acetone xylol (1:1); (4) 1 min in xylol; (5) 2 min in xylol. Mount cover slips using permount while slides are still wet with xylol.

If slides are not stained immediately, store them in slide boxes. It is important in the field to keep slides dry and free from dust. Place a small packet of silica gel crystals in each box and tape the sides of the box to make it airtight. Store slide boxes in plastic bags.

6.2 Collecting Parasites

Mammals are usually hosts of many parasites. Those found externally on the body are ectoparasites; parasites found in internal organs are referred to as endoparasites.

ECTOPARASITES

Ectoparasites are insects (parasitic flies, fly larvae, lice, and fleas) or arachnids (mites and ticks) that feed on the body fluids, dead skin, tissues, or hair of the host mammal. In bats usual parasites are bat flies and mites, whereas lice, mites, ticks, and fleas are commoner on rodents and other mammals. Although many techniques exist to obtain ectoparasites, the following is a simple method suitable for field use. If the mammal is alive, put it in a clean plastic or cloth bag with a piece of cotton soaked in chloroform or ether to kill the ectoparasites and the host. Shake the dead mammal and the inverted bag over a white porcelain tray or a piece of white paper and recover the parasites which will be conspicuously dark on the white background. A vigorous brushing of the fur with a toothbrush will remove any ectoparasites that are caught in the fur. Because ticks imbed in the skin, they are difficult to remove from the host without damaging the mouth parts. For fluid-preserved specimens, leave ticks in position on the mammal and make a note on your data sheets that these ectoparasites are present. For study skins detach the ticks by snipping away some of the skin of the host in which it is imbedded.

There are several precautions to follow when collecting ectoparasites. It is essential to separate different species of mammals in collecting bags. If different species of mammals are put in the same collecting bag, then host data will be meaningless, for most ectoparasites may move from one mammal to another in the bag. Handle ectoparasites carefully, as legs, wings, and other body parts are delicate. Use jeweller's forceps or fine-tipped brushes moistened in alcohol to pick up ectoparasites. Ensure that instruments, killing bags, and pans are washed and clean before processing a new specimen.

Preserve all ectoparasites in small, screw-cap or rubber-stoppered vials with 70 per cent ethanol (do not use formalin, as it hardens specimens). Write the field number of the host and any other pertinent data (e.g., site on body of host) in waterproof ink or pencil on sturdy paper and place it in the vial. Note on your catalogue sheets or field notebook that ectoparasites were preserved for that particular specimen. If you wish further information on parasites, refer to the appropriate publications listed in the bibliography.

Endoparasites

Endoparasites most frequently found in mammals are helminth worms (trematodes, cestodes, and nematodes). Trematodes or flukes are small, flattened worms that occur in the digestive tract, liver, lungs, and other internal organs. Cestodes or tape worms are long, many–segmented worms that live in the intestine of the host; nematodes or round worms are unsegmented worms found in most organs, including muscles. It is beyond the scope of this manual to describe the many special techniques for preserving endoparasites. But for the field collector who may occasionally find these endoparasites and wishes to preserve them for identification, we recommend that they be stored in 70 per cent alcohol. A small amount of glycerine (5 ml in 100 ml of 70% ethanol) will keep parasites pliable and reduce hardening of tissues. Label vials with the field number of the host and note in the catalogue or field notebook that parasites were preserved. For descriptions of special methods for preserving endoparasites, consult references given in the bibliography.

6.3 Tissues for Biochemical Study

Fresh tissues from various organs (heart, kidney, liver, ovary, testes, and muscle) can be preserved in the field by quick freezing.

Remove the desired organs from freshly killed specimens. Cut a 5 to 10 mm cube of tissue from the organ with scissors or a scalpel and wash in physiological saline to remove contaminating blood. The saline solution should be prepared fresh each day and each tissue sample should be rinsed separately. Ideally the tissue should be quick-frozen in liquid nitrogen, wrapped in aluminium foil and stored in dry ice. Label each sample with a field number and code (e.g., H for heart, K for kidney). Liquid nitrogen is suitable for working at a field station facility but may be impractical and dangerous to use in other field situations. If liquid nitrogen cannot be used, wrap tissues in foil, label, and immediately store in dry ice. A small Styrofoam cooler with a tightly fitting lid makes an ideal dry ice chamber (Sudia et al., 1970). Tissue samples will keep for three days in such a container with dry ice without deterioration. Storage time may be increased by replenishing the dry ice every two days. Once samples are brought to the laboratory, they can be stored in a freezer at −70°C for 6 months.

6.4 Blood Samples

Blood samples may be required for biochemical studies of haemoglobin and serum or plasma proteins, blood parasite studies, or immunological studies. Although blood samples taken in the field may be stored temporarily on wet ice and brought to the laboratory for analysis, generally this is not practical. For studies of blood parasites, slides can be prepared directly in the field. Haemoglobin or serum samples obtained for biochemical analyses should be quick-frozen with liquid nitrogen and stored in a Styrofoam cooler with dry ice as described in section 6.3. Once samples are brought to the laboratory, they should be transferred to a freezer where they can be kept for several months.

Mammals should be anaesthetized before blood is taken. Ether is an effective anaesthetic but an overdose may kill the animal. Sudia et al. (1970) recommended carbon dioxide as an anaesthetic. Mammals can be bled from the heart using a 2 or 5 ml disposable syringe. Clean the skin of the area to be punctured with cotton soaked in water and allow to dry. If you require plasma, rinse syringes with heparin to prevent coagulation. For such mammals as mice, use a 25 gauge ⅜ inch or ⅝ inch needle; for mammals to the size of hares, a 23 gauge 1 inch needle; and for mammals the size of foxes or racoons, an 18 gauge 1 ½ inch needle. For small rodents and bats, some workers prefer to take blood from the orbital sinus. Hold the mammal in the left hand with the thumb exerting sufficient pressure behind the eye to cause the eye to bulge out slightly. Insert a 50 or 100 microlitre micro-sampling pipette (Corning Glass Company, Corning, New York, USA) into the posterior corner of the eye and gently rotate it to rupture the capillaries against the bone. Other methods that may be used to obtain blood include venipuncture and skin puncture (Miale, 1972).

Blood from syringes or micro-sampling pipettes should be discharged into vials (Sudia et al., 1970) for freezing and storage. Label each vial with the field number of the specimen. After the blood sample has been taken, the mammal can be humanely killed (see section 5.1) and prepared as a voucher specimen.

Slides of blood smears for parasite studies are easily prepared in the field. However, there are several precautions to follow when preparing blood slides. Use new slides that are either precleaned by the manufacturer or cleaned with soap and water and rinsed in 95 per cent alcohol. Wipe with a lint free cloth before using. If blood is obtained by cutting a vein, pricking tissue, or cutting the tail, discard the first drop from the syringe to get rid of cellular debris. Use the second drop to prepare the blood smear. With blood obtained directly from a syringe inserted into the heart or a vein, it is not necessary to discard this first drop. Heparinized blood is unsuitable for blood smears.

Place a clean microscope slide on a flat surface and put a small drop of fresh blood about 3 cm (1 inch) from the end. The end of a second slide (spreader slide) is placed on the slide in front of the drop of blood. The spreader slide should be maintained at an angle of about 30 degrees. Pull the spreader slide back onto the drop of blood. When the blood has spread the width of the slide, push the spreader slide forward with a fast, steady motion. Keep the slide flat and allow the smear to air dry. Once fixed or dried, slides can be stored in microscope slide boxes. Label each slide with the field number of the specimen. Follow the precautions listed in section 6.1 for keeping slides free of dust. Freshly prepared films always stain best but if the slides cannot be stained promptly, fix your slides in absolute ethanol or methanol. For detailed information on preparing blood smears and blood stains, consult Miale (1972) and Faust et al. (1970).

For most biochemical studies it is essential to separate plasma and red blood cells and best results will be obtained by centrifuging. If electricity is not available, several suitable hand centrifuges are available (Fisher Scientific Company, 711 Forbes Avenue, Pittsburgh, Pennsylvania, USA). Transfer the fresh whole blood samples to centrifuge tubes and centrifuge at 3000 rev/min for 20 min. The buff-coloured layer (plasma) can be separated from the packed cells by aspiration with a micropipette. Transfer the plasma to glass vials or tubes, label, and freeze. If haemoglobin is required, red blood cells should be haemolyzed. Wash the cells once with cold 0.85 per cent saline and then centrifuge at 3000 rev/min for 20 min. Remove the supernatant by aspiration with a micropipette. Add two volumes of distilled water to the packed cells. After mixing thoroughly, freeze and thaw the solution three times. Haemolysates are then centrifuged at 4000 rev/min for 5 min. Transfer the supernatant to vials and quick-freeze with liquid nitrogen.

6.5 Preserving Stomach Contents

Because the food of such small mammals as rodents and insectivores is usually ground completely by the teeth, identification of food items requires microscopic techniques. These usually involve preparation of microscope slides of stomach contents and comparison with reference slides (Drodz, 1975; Williams, 1962). Preparation of these slides can be tedious and time-consuming and the collector may wish to preserve the contents of stomachs in the field and analyse the stomach material at his/her convenience in the laboratory.

Dissect stomachs from freshly killed specimens. If a freezer is available, the simplest method is to place whole stomachs in plastic bags or vials and freeze them. A label with the specimen's field number should be placed in bags or vials. Another method is to remove the contents of stomachs and allow the material to dry. Dried stomach material can be stored in paper envelopes. In the laboratory the desiccated material from stomachs is soaked for 24 h in water and then examined. This technique, however, may not be suitable for humid climates. A third method is to preserve stomachs and their contents in vials with 80 per cent ethanol. Make an incision through the stomach wall to allow the solution to reach stomach contents. Record the field number on a sturdy label with waterproof ink and include it in the vial.

6.6 Preparation of Sperm Slides

For Study with the Light Microscope

Generally, testes from freshly killed specimens are used for sperm studies; however, Hirth (1960) obtained satisfactory results with smears of the cauda epididymis of specimens preserved in 10 per cent formalin. Spermatozoa can be obtained from the cauda epididymis (Fig. 15) or the seminiferous tubules.

The following technique was used by Forman (1968) to study spermatozoa in North American bats. Place the whole testis in a fixing solution consisting of two parts 100 per cent methanol, four parts 95 per cent ethanol, one part acetone, two parts chloroform and one part 100 per cent propionic acid. To permit rapid fixation, cut the testis into 5 to 10 mm squares. Put a short section of seminiferous tubule on a slide with one drop of lactophenol-cotton blue stain. Lactophenol-cotton blue consists of 20 g phenol crystals, 0.05 g cotton blue (Poirrier's Blue, National Aniline Division), 20 ml lactic acid, 40 ml glycerol, and 20 ml distilled water. Dissolve liquids by heating under hot water tap, then add the phenol crystals and cotton blue. Tease the tubule apart to permit spermatozoa to enter the staining solution. Place a cover slip over the slide and seal the edges with balsam.

Genoways (1973) used the following method for comparing spermatozoa in various species of spiny pocket mice of the genus *Liomys*. Remove the epididymis from freshly killed specimens. Take a small amount of fluid containing sperm and suspend in an isotonic solution of sodium citrate (prepared fresh each day). Place a few drops of the suspension on a microscope slide and let dry. Then fix the spermatozoa with a solution of one part glacial acetic acid and four parts absolute methanol for 10 to 15 s. Then stain the sperm slides in a Coplan jar for 30 min with a 0.02 solution of toluidine blue in water.

FOR STUDY WITH THE SCANNING ELECTRON MICROSCOPE (SEM)

Several recent studies have shown that there is great potential for studying the morphology of mammalian spermatozoa with the SEM. If such studies are contemplated, fix the testes in gluteraldehyde in the field. Remove the testes from freshly killed specimens and wash in physiological saline. Place the testes in a vial with a 1 or 2 per cent solution of gluteraldehyde. Gluteraldehyde is sold commercially as a 25 per cent solution in water (J. T. Baker Chemicals, Canadian Laboratory Supplies, Toronto, Canada). The testes should be cut into 5 to 10 mm squares to permit rapid fixation. In the laboratory tissue is washed and centrifuged to remove the fixative and sperm samples are freeze-dried for later examination with the SEM (Gould et al., 1971).

6.7 Fixing Tissues for Histological Study

If histological studies are contemplated, tissues should be fixed from various organs immediately after the specimen is killed. Proper fixation is essential for preventing any post-mortem deterioration in tissues. In section 5 we discussed the use of buffered neutral formalin for fixing entire mammal specimens. Buffered neutral formalin is also an excellent general purpose fixative for histological work and it can be used to fix many different tissues. For certain histological procedures, however, other fixatives may be desired because they penetrate tissues more rapidly than neutralized formalin and may render tissues more easily stained by certain histological dyes. Tissues can be left in some fixatives (e.g., buffered neutralized formalin) for several months; with other fixatives (e.g., Bouin's), tissues must be transferred to alcohol immediately after fixing. Three of the more widely used fixatives are described here. For more information on fixatives and histological methods, consult Luna (1968).

Bouin's Solution

*picric acid, saturated aqueous solution	750 ml
37–40% formalin	250 ml
glacial acetic acid	50 ml

*if not stored in an aqueous solution, picric acid is highly volatile.

Bouin's solution is frequently used for fixing gastrointestinal tracts, reproductive organs, endocrine glands, and brain tissue. It is a rapid fixative and will fix blocks of tissues in 4 to 12 h depending on their size. Once tissues are fixed, however, they must be washed in two or three changes of 40 per cent alcohol for 4 to 6 h to remove all picric acid. If picric acid is not removed, tissue will undergo deleterious changes. After washing tissues, store in 70 per cent alcohol.

Alcohol-Formalin-Acetic Acid Solution (AFA)

37–40% formalin	10 ml
alcohol, 80%	90 ml
glacial acetic acid	5 ml

A good fixative for rapid fixation, AFA solution has been used for reproductive and gastrointestinal tracts. Small pieces of tissue (2 mm thick) will completely fix in 4 to 6 h. AFA is not suitable for tissue storage and the fixed tissue should be transferred to 70 per cent alcohol.

Formalin-Sodium Acetate Solution

37–40% formalin	100 ml
sodium acetate	20 g
tap water	900 ml

This is an excellent fixative in which to store gross blocks of tissue (e.g., whole brains of such small mammals as bats, shrews, or mice).

7. *Shipping Specimens*

7.1 Methods for Shipping

FLUID-PRESERVED SPECIMENS

Once properly fixed, specimens can be shipped "damp packed", that is, only a small amount of 10 per cent formalin is added to the packing material in the container to keep specimens moist in transit. About 0.25 L (0.5 pt) of fluid is sufficient for a container of 3.8 L (1 gal.). Cotton wool or newspaper should be added to each package, both to limit the movement of specimens and to retain the dampness. Wide-mouthed plastic jars are the most suitable shipping containers but if they are not available, plastic bags may be used if they are well sealed. Glass jars are the least desirable type of container but can be used if properly packed in a strong container and insulated with at least 5 cm (2 inches) of wadded paper or similar material on all sides of each jar. If plastic bags are used, it is recommended that the specimens, cotton wool, and fluid be placed in one bag and sealed by tying the top of the bag; be certain to remove excess air from the bag. Then place the sealed bag inside a second bag, which is in turn sealed by careful tying. The double plastic bag should then be placed in a light, strong container. A tin can with a lid is ideal. If the tin is too large, fill the remaining space with cotton wool or crushed paper. The container should be large enough to absorb the fluid should the bags leak. Seal tin cans with adhesive tape to ensure they will not open or leak in transit. Generally, packages should be kept small enough so that they may be shipped by ordinary parcel post.

Skins and Skulls

It is advisable to ship study skins in small, strong containers, preferably plywood boxes, although strong cardboard boxes may be used. Begin packing by placing a layer of about 5 cm (2 inches) of cotton wool or similar material on the bottom, followed by a layer of dried skins. Add another layer of cotton wool and repeat the process with another layer of skins, until the box is almost full. Allow a top layer of about 5 cm (2 inches) of cotton wool.

If thoroughly dried and free of insects and insect eggs, skulls may be packed in the same box with skins. Otherwise, ship them in a separate strong container that cannot be smashed or crushed under normal shipping conditions. If the skins are likely to be in transit for longer than a few days, add moth balls (paradichlorobenzine) to the container of skins, to protect them from moths or dermestid beetle eggs. Skulls and skeletons should not be placed together in a container with moth balls because the chemical released may inhibit the activity of dermestid beetles that are used by most museums to clean skulls and skeletons.

Skeletons

Skeletons must be dried before shipping. You can pack small skeletons in the same container used for the skins if moth crystals have not been added to the container. Place the skeletons on the bottom of the container, then cover with alternate layers of cotton and skins. Heavy bones from large mammals should be packed separately in sturdy wooden crates. If a large skeleton is properly dried, place it in a plastic bag just before shipping to reduce the odour. Collectors shipping large skeletons should notify the receiving museum prior to shipping.

Data

Pages of the original field catalogue, field notes, and any pertinent topographic maps should be mailed separately from specimens as first class or air mail and registered if from a country where mail service is inadequate.

7.2 Import/Export Regulations

Convention on International Trade in Endangered Species

Canada, the United States, and approximately 40 other countries have signed a Convention on International Trade in Endangered Species. The Convention prohibits all imports and exports of protected species except under permit. These regulations apply to all international shipments and, in addition to living mammals, to parts and derivatives of endangered species. Permit requirements must be observed for any shipment to or from a member country, even if the other country involved is not a member of the Convention. Species protected are listed in three Convention appendices. Appendix I of the Convention lists species threatened with extinction; Appendix II lists species that must be monitored to avoid the threat of extinction; and Appendix III lists species placed there by individual countries to reinforce domestic conservation measures. All shipments of species listed in Appendix I of the Convention require two permits—one from the importing country and another from the exporting country. Export permits must be issued from the country of origin for species listed in Appendix II. International shipments of Appendix III species require either an export certificate from the country that listed the species or a certificate of origin from any country.

Shipments to Canada

All shipments to Canadian museums should be labelled with a declaration form that indicates the total number of specimens, the general kind of specimens (bats or rodents), and that specimens are preserved (e.g., 100 preserved bats) rather than live or frozen. Shipments of living mammals must have Canadian Department of Agriculture, Health of Animal Branch permits.

The shipment should also be marked "No Commercial Value (NCV), For Scientific Research, No Endangered Species". If the collector insures the shipment for more than $150.00, he must complete special Canada Customs invoice forms for shipments from outside Canada; these invoice forms are available from museums upon request. Shipments should also be labelled "In Bond to Destination" which ensures that the shipment can only be inspected by customs officials at the destination and not at border points. Notify the museum of a forthcoming shipment and, when sending specimens by air freight, forward a copy of the Airway Bill.

Permits to import or export species listed under the Convention on International Trade in Endangered Species may be obtained from the Canadian Wildlife Service in Ottawa or any provincial or territorial wildlife headquarters.

Shipments to the United States

Packages should be clearly marked on the outside with the name and address of the shipper and of the consignee and give an accurate statement of the contents (species and numbers of each species). Label the package "Scientific Specimens; No Endangered Species; No Commercial Value".

DESIGNATED PORTS OF ENTRY

Nonendangered mammals and parts thereof from countries other than Canada may enter or leave the US without a permit only through eight designated ports of entry: Los Angeles, San Francisco, Miami, Honolulu, Chicago, New Orleans, New York, and Seattle. Species protected under the Convention on International Trade in Endangered Species and the US Endangered Species Act must enter the US through these same eight cities. Mammals other than endangered species may be imported, for final destination only, into Alaska through Anchorage, Fairbanks, Juneau, or Tok Junction; or into either Puerto Rico or the Virgin Islands through San Juan, Puerto Rico.

Specimens obtained legally in Canada may enter the USA without permit through any of 25 border points if no endangered species are included. Contact the Division of Law Enforcement, US Fish & Wildlife Service, Washington, D.C., USA for a list of these border points. If no endangered species are included, mammals collected legally in Mexico may enter the US without permit through any of seven border points: Calexico or San Diego–San Ysidro, California; Nogales or San Luis, Arizona; Brownsville, El Paso, or Laredo, Texas.

Scientific specimens of nonendangered species may enter or leave the US at nondesignated ports of entry under permit.

PERMITS

To import specimens through the designated ports of entry previously given, the collector needs a valid collecting permit and an export permit (if required by the country of origin) and a Declaration of Importation of Fish & Wildlife (Form 3-177). You must file copies of these documents with the District Director of Customs at the port of entry.

For importation of mammal specimens through nondesignated ports of entry, you must satisfy the previously given requirements and have a permit for the nondesignated port from the U.S. Fish & Wildlife Service.

Permits for importing specimens of mammals listed in the Convention on International Trade in Endangered Species of the federal Endangered Species Act can be obtained from the Federal Wildlife Permit Office, U.S. Fish & Wildlife Service, Washington, DC 20240, USA. Note that although some species of mammals are listed in both the convention and the Endangered Species Act, the two lists do not contain identical species.

A permit is required to import living material, including tissue cultures, cell lines, and blood and serum that could serve as a vector for pathogenic organisms. Apply to the U.S. Department of Agriculture, Animal and Plant Health Inspection Service, Veterinary Services, Washington, D.C., USA for this permit.

INTERSTATE TRANSPORT OF MAMMAL SPECIMENS

Packages or containers in which specimens are transported must be clearly marked on the outside with the same information as described for shipments from outside the US. Because it is illegal under the Lacey Act to import or ship in interstate commerce any wildlife taken in violation of state or local laws, scientific collectors must familiarize themselves with appropriate laws for the state or region concerned.

8. *Public Health Hazards*

It is beyond the scope of this manual to discuss the numerous potentially dangerous diseases that are carried by wild mammals. Several useful references are given in the bibliography. Because diseases are often local in their distribution, it is impossible to list specific diseases that the collector may encounter. As a result, you are advised to familiarize yourself with the disease hazards that may be present in your local area. Consult your physician about vaccinations that may be taken against diseases in your area, for example, rabies, plague. All collectors should be vaccinated against tetanus.

Take care to avoid being bitten when handling live mammals and use leather gloves to protect the hands. If you are collecting species that are susceptible to rabies (bats, foxes, mongooses, skunks), regard all captures as potentially rabid. Some common sense precautions to reduce health hazards can be followed when preparing specimens. Wear rubber gloves when dissecting mammals and, for additional protection, use disposable paper face masks. Avoid contact with urine and faeces, which often contain infectious agents. Dissecting instruments should be cleaned and disinfected with Dettol or a 3 per cent phenol solution after use. Immediately wash any cuts or abrasions with soap and treat with an antiseptic. Use similar precautions when handling road kills. If symptoms such as chronic respiratory distress, influenza-like sickness, swelling of lymph nodes, high temperature, vomiting, or diarrhoea occur in conjunction with the handling of specimens, they should be regarded with suspicion and medical advice should be sought.

9. *Acknowledgements*

We gratefully acknowledge the assistance and encouragement of staff of the Department of Mammalogy, ROM, during the preparation of the manuscript. The following staff deserve special thanks: J. R. Tamsitt for his advice on special techniques and for critically reviewing the manuscript, Judith Eger for critically reviewing the manuscript, Sophie Poray for drawing some of the illustrations, and Nancy Grepe for typing the numerous revisions of the manuscript. Other illustrations were done by Anker Odum and Julian Mulock, Exhibit Design Services, ROM. The cover drawing is by Peter Buerschaper, Exhibit Design Services, ROM. Many of the techniques described in the manual for preparing study skins were developed by John G. Williams.

10. *Selected Bibliography*

General Reference

ANDERSON, R. M.
1965 Methods of collecting and preserving vertebrate animals. 4th ed. rev. National Museum of Canada, Bulletin 69:1–199.

BRITISH MUSEUM (NATURAL HISTORY)
1968 Instructions for collectors no. 1. Mammals (non-marine). 2nd ed. British Museum (Natural History) Publication 665:1–55.

BROWN, J. C. and D. M. STODDART
1977 Killing mammals and general post-mortem methods. Mammal Review 7:63–94.

DEBLASE, A. F. and R. E. MARTIN
1974 A manual of mammalogy with keys to the families of the world. Dubuque, W. C. Brown. 329 pp.

GILES, R.H., Jr., ed.
1971 Wildlife management techniques. 3rd ed. rev. Washington, D.C., The Wildlife Society. 633 pp.

HALL, E. R.
1962 Collecting and preparing study specimens of vertebrates. University of Kansas, Museum of Natural History, Miscellaneous Publication 30:1–46.

KNUDSEN, J. W.
1966 Biological techniques; collecting, preserving, and illustrating plants and animals. New York, Harper & Row. 525 pp.

PETERSON, R. L.
1965 Collecting bat specimens for scientific purposes. Toronto, Dept. of Mammalogy, Royal Ontario Museum. 8 pp.

WILLIAMS, S.L., R. LAUBACK, and H. H. GENOWAYS
1977 A guide to the management of recent mammal collections. Carnegie Museum of Natural History, Special Publication 4:1–106.

Equipment

DOWLER, R. C. and H. H. GENOWAYS
1976 Supplies and suppliers for vertebrate collections. Museology, Texas Tech University, 4:1–83.

Locality Data

AXTELL, R. W.
1965 More on locality data and its presentation. Systematic Zoology 14:64–66.

REIMER, W. J.
1954 Formulation of locality data. Systematic Zoology 3:138–140.

Mammal Collecting

BAKER, R. J. and S. L. WILLIAMS
1972 A live trap for pocket gophers. Journal of Wildlife Management 36:1320–1322.

GREENHALL, A. M. and J. L. PARADISO
1968 Bats and bat banding. United States Dept. of the Interior, Bureau of Sport Fisheries and Wildlife, Resource Publication 72:1–47.

MANITOBA DEPT. OF RENEWABLE RESOURCES AND TRANSPORTATION SERVICE
1965 The Manitoba trappers' guide. Rev. ed. Winnipeg, Dept. of Renewable Resources and Transportation Service. 122 pp. [good review of trapping and skinning methods]

NELLES, C. H. and R. D. TABER
1974 A conical pitfall trap for small mammals. Northwest Science 48:102–103.

STAINS, H. J.
1962 Game biology and game management: a laboratory manual. Minneapolis, Burgess. 143 pp. [trapping methods]

TIDEMAN, C. R. and D. P. WOODSIDE
1978 A collapsible bat-trap and a comparison of results obtained with trap and with mist nets. Austrialian Wildlife Research 5:355–362.

TUTTLE, M. D.
1974 An improved trap for bats. Journal of Mammalogy 55:475–477.
1976 Collecting techniques. *In* Baker, R.J. et al., eds., Biology of bats of the New World family Phyllostomatidae Part I. Texas Tech University, Musuem, Special Publications 10:71–88.

TWIGG, G. I.
1975 Catching mammals. Mammal Review 5:83–100.

WEINER, J. G. and H. G. SMITH
1972 Relative efficiencies of four small mammal traps. Journal of Mammalogy 53:868–873.

WINGATE, L. R. and J. MEESTER
1977 A field test of six types of live-trap for african rodents. Zoologica Africana 12:215–223.

YATES, T. L. and D. J. SCHMIDLY
1975 Karyotype of the eastern mole (*Scalopus aquaticus*), with comments on the karyology of the family Talpidae. Journal of Mammalogy 56:902–905. [description of a live trap for moles]

Preparation

DOWNING, S. C.
1945 Color changes in mammal skins during preparation. Journal of Mammalogy 26:128–132.

QUAY, W. M.
1974 Bird and mammal specimens in fluid—objectives and methods. Curator 17:91–104.

SETZER, H. W.
1968 Directions for preserving mammals for museum study. United States National Museum, Smithsonian Institution, Information Leaflet 380:1–19.

SMITH, D. A.
1968 On the use of tapered wires for study skins of small mammals. Journal of Mammalogy 49:787–790.

Public Health

IRVIN, A. D. and J. E. COOPER
1972 Possible health hazards associated with the collection and handling of post-mortem zoological material. Mammal Review 2:43–54.

McDIARMID, A., ed.
1969 Diseases in free-living wild animals. Zoological Society of London, Symposium 24. London, Academic Press. 332 pp.

Regulations

ASSOCIATION OF SYSTEMATICS COLLECTIONS
1977 Index to U.S. federal wildlife regulations. Lawrence, University of Kansas, Museum of Natural History. 278 pp.

CANADIAN WILDLIFE SERVICE
1973 Convention on international trade in endangered species of wild flora and fauna. Canadian Wildlife Service reprint. Ottawa, Environment Canada. 20 pp.

EDWARDS, S. R. and L. D. GROTTA, eds.
1976 Systematics collections and the law. Lawrence, Association of Systematics Collections. 29 pp.

GENOWAYS, H. H. and J. R. CHOATE
1976 Federal regulations pertaining to collection, import, export and transport of scientific specimens of mammals. Journal of Mammalogy 57(2) suppl.: 1–9.

McGAUGH, M. H. and H. H. GENOWAYS
1976 State laws as they pertain to scientific collecting permits. Museology, Texas Tech University, 2:1–81.

SINGLETON, M.
1977 Endangered species legislation in Canada. *In* Mosqin, T. and C. Suchal, eds., Canada's threatened species and habitats. Canadian Nature Federation, Special Publication 6:19–21.

Reproductive Data

CONWAY, C. H.
1955 Embryo resorption and placental scar formation in the rat. Journal of Mammalogy 36:516–532.

ECKSTEIN, P. and S. ZUCKERMAN
1956 Morphology of the reproductive tract. *In* Parkes, A. S., ed., Marshall's physiology of reproduction, vol. 1, pt. 1. London, Longman Green, pp. 43–155.

JAMESON, E. W., Jr.
1950 Determining fecundity in male small mammals. Journal of Mammalogy 31:433–436.

ROLAN, R. G. and H. T. GIER
1967 Correlation of embryo and placental scar counts of *Peromyscus maniculatus* and *Microtus ochrogaster*. Journal of Mammalogy 48:317–319.

Special Techniques

BAKER, R. J.
1970 The role of karyotypes in phylogenetic studies of bats. *In* Slaughter, B. H. and D. W. Walton, eds., About bats: a chiropteran biology symposium. Dallas, Southern Methodist University Press, pp. 303–312.

DRODZ, A.
1975 Analysis of stomach contents of small mammals. *In* Grodzinski, W., R. Z. Klekowski, and A. Duncan, eds., Methods for ecological bioenergetics. IBP handbook 24. Oxford, Blackwell Scientific, pp. 337–341.

FAUST, E. C., P. F. RUSSELL, and A. C. JUNG
1970 Craig and Faust's clinical parasitology. 8th ed. Philadelphia, Lea and Febiger. 890 pp.

FORMAN, G. L.
1968 Comparative gross morphology of spermatozoa of two families of North American bats. University of Kansas Science Bulletin 16:901–928.

GENOWAYS, H. H.
1973 Systematics and evolutionary relationships of spiny pocket mice, genus *Liomys*. Texas Tech University, Museum, Special Publications 5:1–292. [technique for preparing rodent sperm slides]

GOULD, K. G., L. J. D. ZANEVELD, and W. L. WILLIAMS
1971 Mammalian gametes: a study with the scanning electron microscope. Chicago, Proceedings of the Fourth Annual Scanning Electron Microscope Symposium, part 1, pp. 289–295.

HIRTH, M. F.
1960 The spermatozoa of some North American bats and rodents. Journal of Morphology 106:77–83.

KAISER, M. N. and H. HOOGSTRAAL
1968 Simple field techniques and laboratory method for recovering living ticks (Ixoidea) from hosts. Journal of Parasitology 54:188–189.

LUNA, L. G.
1968 Manual of histological staining methods of the Armed Forces Institute of Pathology. 3rd ed. New York, McGraw Hill. 258 pp.

MEYER, M. C. and O. W. OLSEN
1971 Essentials of parasitology. Dubuque, W. C. Brown. 305 pp.

MIALE, J. B.
1972 Laboratory medicine: hematology. 4th ed. Saint Louis, C. V. Mosby. 1318 pp.

SELLER, M. J.
1971 Direct and repeatable bone marrow chromosome preparations from living mice. Stain Technology 46:285–288.

SUDIA, W. D., R. D. LORD, and R. O. HAYES
1970 Collection and processing of vertebrate specimens for arbovirus studies. Washington, United States Dept. of Health, Education, and Welfare, Public Health Service. 65 pp.

WATSON, G. E. and A. B. AMERSON, Jr.
1967 Instructions for collecting bird parasites. United States National Museum, Smithsonian Institution, Information Leaflet 477:1–12.

WILLIAMS, O.
1962 A technique for studying microtine food habits. Journal of Mammalogy 43:365–368.

11. Appendices

11.1 Appendix 1

CHECKLIST OF FIELD EQUIPMENT

For a listing of suppliers of field equipment, consult Dowler and Genoways (1976).

Collecting Bats

mist nets—12.5 m × 2 m (42 ft × 7 ft), 9 m × 2 m (30 ft × 7 ft)
hand net
machete—for cutting poles
headlamp and batteries—for working in caves or checking nets at night
flashlight
long forceps (25 cm; 10 inches)—for collecting bats in attics or rock crevices
cloth collecting bags
leather gloves—for handling live specimens

Collecting Other Mammals

museum special snap traps
Victor rat-traps
Sherman live traps
juice or coffee cans—for pitfall traps
cloth collecting bags
bait—peanut butter, rolled oats, dried fruits, nuts
coloured, plastic flagging tape—for marking trap sites
compass and topographic maps of study areas
tobacco tin or plastic jar—for carrying bait
knapsack—for carrying traps
cord, wire—for tying traps to stakes
guns, ammunition
leather gloves
cotton wool—nesting material for live traps

Measuring and Preparing Specimens

weight scales (metric)—Pesola spring balances
metal tape-measure (metric), plastic ruler (metric)
catalogue sheets
field tags, waterproof paper
hand lens—for examining external genitalia and internal sex organs
notebook—for field notes
pencils, pen, and waterproof ink

small waterproof tags—for labelling vials
small vials (1–4 dram size)—for embryos and parasites
borax or magnesium carbonate powder
gauze or cheesecloth bags—for skulls and skeletons
disinfectant (Dettol or 3% phenol)—for sterilizing dissecting instruments
sawdust or cornmeal
toothbrush—for brushing fur
disposable gloves
chloroform or ether—for killing small mammals
Euthanyl (sodium pentobarbital)—for killing larger mammals
70 per cent ethanol—for ectoparasites, cleaning blood off fur
40 per cent formaldehyde, salts for neutralizing formalin
hypodermic syringe and needles—for injecting specimens
plastic bags, various sizes
moth balls (paradichlorobenzene crystals)
cotton—for stuffing study skins
wire for tails of skins—no. 12, 16, 18, 20, 22, 24, 25 gauge (preferably Monel wire)
glass-headed pins, various sizes
needles, various shapes and sizes
pinning board, corkboard, or wallboard
assortment of cardboard or art board—for flat skins
white linen thread, sizes 40, 30, 25
white cotton thread, sizes 50, 60
pliers, wire cutters
drying chest—for storing drying skins
scalpel handle, no. 3 and no. 4 sizes
scalpel blades, no. 20 size (for no. 4 handle) and no. 10 size (for no. 3 handle)
scissors, one small pair with sharp points, one large pair
forceps, one pair with fine tips, one pair with blunt tips
bone cutters
skinning knife—for large mammals
sharpening stone
table salt (NaCl)—for drying large skins
burlap sacks—for skulls and bones of large mammals
masking tape
electrical or adhesive tape—for securing lids on glass vials and taping lids of mailing containers

11.2 Appendix 2

Karyotype Kit for 120 Small Mammals

Sodium citrate 15 packages of 0.5 g quantity
5 packages of 0.75 g quantity
1 hand centrifuge
24-15 ml glass centrifuge tubes
24-1 ml tuberculin syringes

2-2 ml Luer Lock syringes
4-5 ml Luer Lock syringes
1 doz. no. 26 syringe needles (½ inch)
1 doz. no. 24 syringe needles (1 inch)
1 doz. no. 21 syringe needles (1 ½ inches)
250 (1 box) disposable glass pipettes (5.75 inches)
48 rubber pipette bulbs
1 plastic centrifuge rack
2 centrifuge tube cleaning brushes
5 gross (720) precleaned microscope slides with frosted ends (75 mm × 25 mm size)
3 pints absolute methanol
2 pints glacial acetic acid
4-10 mg vials Velbe
2-100 ml serum bottles of bacteriostatic sodium chloride
1-125 ml nalgene Erlenmeyer flask—for sodium citrate solution
9 slide boxes (100 slide capacity)
1-100 ml nagrene Erlenmeyer flask—for fixative
1-50 ml nagrene graduated cylinder for measuring fixative
matches
silica gel crystals
grease pencil—for labelling centrifuge tubes
100 ml distilled water

For Staining

500 ml xylol
500 ml acetone
100 ml Giemsa stain
100 ml permount
5-1 oz. boxes cover slips (no. 1, 24 mm × 50 mm)
6 Coplan jars
1-50 ml polypropylene graduated cylinder—for measuring stain
1 polypropylene funnel, filter paper—for filtering stain